Collins

ROYAL
OBSERVATORY
GREENWICH

2023 GUIDE
to the
NIGHT SKY

Storm Dunlop and Wil Tirion

Published by Collins
An imprint of HarperCollins*Publishers*
Westerhill Road, Bishopbriggs, Glasgow G64 2QT
www.harpercollins.co.uk

HarperCollins*Publishers*
1st Floor, Watermarque Building, Ringsend Road, Dublin 4, Ireland

In association with
Royal Museums Greenwich, the group name for the National Maritime Museum,
Royal Observatory Greenwich, Queen's House and *Cutty Sark*
www.rmg.co.uk

A catalogue record for this book is available from the British Library

ISBN 978-0-00-839354-0

10 9 8 7 6 5 4 3 2 1

Printed in Italy by Rotolito S.p.A.

If you would like to comment on any aspect of this book, please contact us at the above address or online.
e-mail: collinsmaps@harpercollins.co.uk

 facebook.com/CollinsAstronomy

@CollinsAstro

MIX
Paper from
responsible sources
FSC www.fsc.org **FSC™ C007454**

This book is produced from independently certified
FSC™ paper to ensure responsible forest management.

For more information visit: www.harpercollins.co.uk/green

Contents

Introduction

The aim of this Guide is to help people find their way around the night sky, by showing how the stars that are visible change from month to month and by including details of various events that occur throughout the year. The objects and events described may be observed with the naked eye, or nothing more complicated than a pair of binoculars.

The conditions for observing naturally vary over the course of the year. During the summer, twilight may persist throughout the night and make it difficult to see the faintest stars. There are three recognized stages of twilight: civil twilight, when the Sun is less than 6° below the horizon; nautical twilight, when the Sun is between 6° and 12° below the horizon; and astronomical twilight, when the Sun is between 12° and 18° below the horizon. Full darkness occurs only when the Sun is more than 18° below the horizon. During nautical twilight, only the very brightest stars are visible. During astronomical twilight, the faintest stars visible to the naked eye may be seen directly overhead, but are lost at lower altitudes. As the diagram shows, during June and most of July full darkness never occurs

at the latitude of London, and at Edinburgh nautical twilight persists throughout the whole night, so at that latitude only the very brightest stars are visible.

Another factor that affects the visibility of objects is the amount of moonlight in the sky. At Full Moon, it may be very difficult to see some of the fainter stars and objects, and even when the Moon is at a smaller phase it may seriously interfere with visibility if it is near the stars or planets in which you are interested. A full lunar calendar is given for each month and may be used to see when nights are likely to be darkest and best for observation.

The celestial sphere

All the objects in the sky (including the Sun, Moon and stars) appear to lie at some indeterminate distance on a large sphere, centred on the Earth. This **celestial sphere** has various reference points and features that are related to those of the Earth. If the Earth's rotational axis is extended, for example, it points to the North and South Celestial Poles, which are thus in line with the North and South Poles on Earth. Similarly, the **celestial**

The duration of twilight throughout the year at London and Edinburgh.

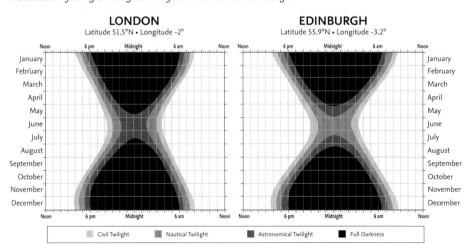

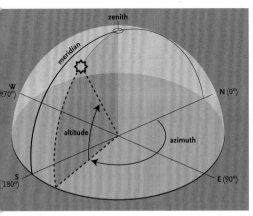

Measuring altitude and azimuth on the celestial sphere.

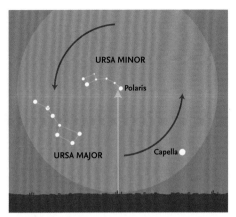

The altitude of the North Celestial Pole equals the observer's latitude.

equator lies in the same plane as the Earth's equator, and divides the sky into northern and southern hemispheres. Because this Guide is written for use in Britain and Ireland, the area of the sky that it describes includes the whole of the northern celestial hemisphere and those portions of the southern that become visible at different times of the year. Stars in the far south, however, remain invisible throughout the year, and are not included.

It is useful to know some of the special terms for various parts of the sky. As seen by an observer, half of the celestial sphere is invisible, below the horizon. The point directly overhead is known as the **zenith**, and the (invisible) one below one's feet as the **nadir**. The line running from the north point on the horizon, up through the zenith and then down to the south point is the **meridian**. This is an important invisible line in the sky, because objects are highest in the sky, and thus easiest to see, when they cross the meridian in the south. Objects are said to **transit**, when they cross this line in the sky.

In this book, reference is frequently made in the text and in the diagrams to the standard compass points around the horizon. The position of any object in the sky may be described by its **altitude** (measured in degrees above the horizon), and its **azimuth** (measured in degrees from north 0°, through east 90°,

south 180° and west 270°). Experienced amateurs and professional astronomers also use another system of specifying locations on the celestial sphere, but that need not concern us here, where the simpler method will suffice.

The celestial sphere appears to rotate about an invisible axis, running between the North and South Celestial Poles. The location (i.e., the altitude) of the Celestial Poles depends entirely on the observer's position on Earth or, more specifically, their latitude. The charts in this book are produced for the latitude of 50°N, so the North Celestial Pole (NCP) is 50° above the northern horizon. The fact that the NCP is fixed relative to the horizon means that all the stars within 50° of the pole are always above the horizon and may, therefore, always be seen at night, regardless of the time of year. This northern circumpolar region is an ideal place to begin learning the sky, and ways to identify the circumpolar stars and constellations will be described shortly.

The ecliptic and the zodiac

Another important line on the celestial sphere is the Sun's apparent path against the background stars – in reality the result of the Earth's orbit around the Sun. This is known as the **ecliptic**. The point where the Sun, apparently moving along the ecliptic, crosses the celestial equator from south to north is

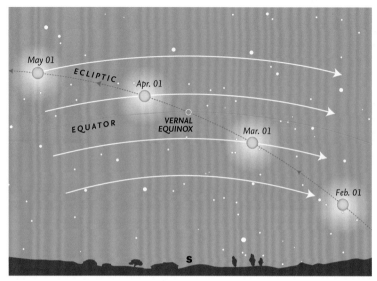

The Sun crossing the celestial equator in spring.

known as the vernal (or spring) equinox, which occurs on March 20. At this time (and at the autumnal equinox, on September 22 or 23, when the Sun crosses the celestial equator from north to south) day and night are almost exactly equal in length. (There is a slight difference, but that need not concern us here.) The vernal equinox is currently located in the constellation of Pisces, and is important in astronomy because it defines the zero point for a system of celestial coordinates, which is, however, not used in this Guide.

The Moon and planets are to be found in a band of sky that extends 8° on either side of the ecliptic. This is because the orbits of the Moon and planets are inclined at various angles to the ecliptic (i.e., to the plane of the Earth's orbit). This band of sky is known as the zodiac and, when originally devised, consisted of twelve **constellations**, all of which were considered to be exactly 30° wide. When the constellation boundaries were formally established by the International Astronomical Union in 1930, the exact extent of most constellations was altered and, nowadays, the ecliptic passes through thirteen constellations.

Because of the boundary changes, the Moon and planets may actually pass through several other constellations that are adjacent to the original twelve.

The constellations

Since ancient times, the celestial sphere has been divided into various constellations, most dating back to antiquity and usually associated with certain myths or legendary people and animals. Nowadays, the boundaries of the constellations have been fixed by international agreement and their names (in Latin) are largely derived from Greek or Roman originals. Some of the names of the most prominent stars are of Greek or Roman origin, but many are derived from Arabic names. Many bright stars have no individual names and, for many years, stars were identified by terms such as 'the star in Hercules' right foot'. A more sensible scheme was introduced by the German astronomer Johannes Bayer in the early seventeenth century. Following his scheme – which is still used today – most of the brightest stars are identified by a Greek letter followed by the genitive form of the constellation's Latin name. An example is the Pole Star, also known as Polaris and α Ursae Minoris (abbreviated α UMi). The Greek alphabet is shown on page 109 with a list of all the constellations that may be seen from latitude 50°N, together with abbreviations, their genitive forms and English names. Other naming schemes exist for fainter stars, but are not used in this book.

Asterisms

Apart from the constellations (88 of which cover the whole sky), certain groups of stars, which may form a part of a larger constellation or cross several constellations, are readily recognizable and have been given individual names. These groups are known as **asterisms**, and the most famous (and well-known) is the 'Plough', the common name for the seven brightest stars in the constellation of Ursa Major, the Great Bear. The names and details of some asterisms mentioned in this book are given in the list on page 110.

Magnitudes

The brightness of a star, planet or other body is frequently given in magnitudes (mag.). This is a mathematically defined scale where larger numbers indicate a fainter object. The scale extends beyond the zero point to negative numbers for very bright objects. (Sirius, the brightest star in the sky is mag. -1.4.) Most observers are able to see stars to about mag. 6, under very clear skies.

The Moon

Although the daily rotation of the Earth carries the sky from east to west, the Moon gradually moves eastwards by approximately its diameter (about half a degree) in an hour. Normally, in its orbit around the Earth, the Moon passes above or below the direct line between Earth and Sun (at New Moon) or outside the area obscured by the Earth's shadow (at Full Moon). Occasionally, however, the three bodies are more-or-less perfectly aligned to give an **eclipse**: a solar eclipse at New Moon or a lunar eclipse at Full Moon. Depending on the exact circumstances, a solar eclipse may be merely partial (when the Moon does not cover the whole of the Sun's disk); annular (when the Moon is too far from Earth in its orbit to appear large enough to hide the whole of the Sun); or total. Total and annular eclipses are visible from very restricted areas of the Earth, but partial eclipses are normally visible over a wider area.

Somewhat similarly, at a lunar eclipse, the Moon may pass through the outer zone of the Earth's shadow, the **penumbra** (in a penumbral eclipse, which is not generally perceptible to the naked eye), so that just part of the Moon is within the darkest part of the Earth's shadow, the **umbra** (in a partial eclipse); or completely within the umbra (in a total eclipse). Unlike solar eclipses, lunar eclipses are visible from large areas of the Earth.

Occasionally, as it moves across the sky, the Moon passes between the Earth and individual planets or distant stars, giving rise to an **occultation**. As with solar eclipses, such occultations are visible from restricted areas of the world.

The planets

Because the planets are always moving against the background stars, they are treated in some detail in the monthly pages and information is given when they are close to other planets, the Moon or any of five bright stars that lie near the ecliptic. Such events are known as **appulses** or, more frequently, as **conjunctions**. (There are technical differences in the way these terms are defined – and should be used – in astronomy, but these need not concern us here.) The positions of the planets are shown for every month on a special chart of the ecliptic.

The term conjunction is also used when a planet is either directly behind or in front of the Sun, as seen from Earth. (Under normal circumstances it will then be invisible.) The conditions of most favourable visibility depend on whether the planet is one of the two known as **inferior planets** (Mercury and Venus) or one of the three **superior planets** (Mars, Jupiter and

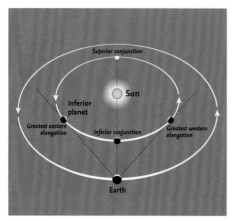

Inferior planet.

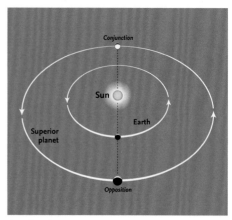

Superior planet.

Saturn) that are covered in detail. (Some details of the fainter superior planets, Uranus and Neptune, are included in this Guide, and special charts are given on page 25.)

The inferior planets are most readily seen at eastern or western **elongation**, when their angular distance from the Sun is greatest. For superior planets, they are best seen at **opposition**, when they are directly opposite the Sun in the sky, and cross the meridian at local midnight.

It is often useful to be able to estimate angles on the sky, and approximate values may be obtained by holding one hand at arm's length. The various angles are shown in the diagram, together with the separations of the various stars in the Plough.

Meteors

At some time or other, nearly everyone has seen a **meteor** – a 'shooting star' – as it flashed across the sky. The particles that cause meteors – known technically as 'meteoroids' – range in size from that of a grain of sand (or even smaller) to the size of a pea. On any night of the year there are occasional meteors, known as **sporadics**, that may travel in any direction. These occur at a rate that is normally between three and eight in an hour. Far more important, however, are **meteor showers**, which occur at fixed periods of the year, when the

Earth encounters a trail of particles left behind by a comet or, very occasionally, by a minor planet (asteroid). Meteors always appear to diverge from a single point on the sky, known as the **radiant**, and the radiants of major showers are shown on the charts. Meteors that come from a circular area 8° in diameter around the radiant are classed as belonging to the particular shower. All others that do not come from that area are sporadics (or, occasionally from another shower that is active at the same time). A list of the major meteor showers is given on page 31.

Although the positions of the various shower radiants are shown on the charts, looking directly at the radiant is not the most effective way of seeing meteors. They are most likely to be noticed if one is looking about 40–45° away from the radiant position. (This is approximately two hand-spans as shown in the diagram for measuring angles.)

Other objects

Certain other objects may be seen with the naked eye under good conditions. Some were given names in antiquity – Praesepe is one example – but many are known by what are called 'Messier numbers', the numbers in a catalogue of nebulous objects compiled by Charles Messier in the late eighteenth century. Some, such as the Andromeda Galaxy, M31, and the Orion Nebula, M42, may be seen

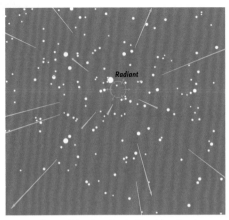

Meteor shower (showing the April Lyrid radiant).

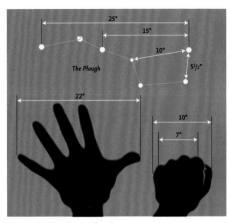

Measuring angles in the sky.

by the naked eye, but all those given in the list will benefit from the use of binoculars. Apart from galaxies, such as M31, which contain thousands of millions of stars, there are also two types of cluster: open clusters, such as M45, the Pleiades, which may consist of a few dozen to some hundreds of stars; and globular clusters, such as M13 in Hercules, which are spherical concentrations of many thousands of stars. One or two gaseous nebulae, consisting of gas illuminated by stars within them, are also visible. The Orion Nebula, M42, is one, and is illuminated by the group of four stars, known as the Trapezium, which may be seen within it by using a good pair of binoculars.

Some interesting objects.

Messier / NGC	Name	Type	Constellation	Maps (months)
—	Hyades	open cluster	Taurus	Sep. – Mar.
—	Double Cluster	open cluster	Perseus	All year
—	Melotte 111 (Coma Cluster)	open cluster	Coma Berenices	Jan. – Aug.
M3	—	globular cluster	Canes Venatici	Jan. – Sep.
M4	—	globular cluster	Scorpius	May – Aug.
M8	Lagoon Nebula	gaseous nebula	Sagittarius	Jun. – Aug.
M11	Wild Duck Cluster	open cluster	Scutum	May – Oct.
M13	Hercules Cluster	globular cluster	Hercules	Feb. – Nov.
M15	—	globular cluster	Pegasus	Jun. – Dec.
M20	Trifid Nebula	gaseous nebula	Sagittarius	Jun. – Aug.
M22	—	globular cluster	Sagittarius	Jun. – Sep.
M27	Dumbbell Nebula	planetary nebula	Vulpecula	May – Dec.
M31	Andromeda Galaxy	galaxy	Andromeda	All year
M35	—	open cluster	Gemini	Oct. – May
M42	Orion Nebula	gaseous nebula	Orion	Nov. – Mar.
M44	Praesepe	open cluster	Cancer	Nov. – Jun.
M45	Pleiades	open cluster	Taurus	Aug. – Apr.
M57	Ring Nebula	planetary nebula	Lyra	Apr. – Dec.
M67	—	open cluster	Cancer	Dec. – May
NGC 752	—	open cluster	Andromeda	Jul. – Mar.
NGC 3242	Ghost of Jupiter	planetary nebula	Hydra	Feb. – May

The Northern Circumpolar Constellations

The northern circumpolar stars are the key to starting to identify the constellations. For anyone in the northern hemisphere they are visible at any time of the year, and nearly everyone is familiar with the seven stars of the Plough – known as the Big Dipper in North America – an asterism that forms part of the large constellation of **Ursa Major** (the Great Bear).

Ursa Major

Because of the movement of the stars caused by the passage of the seasons, Ursa Major lies in different parts of the evening sky at different periods of the year. The diagram below shows its position at the beginning of the four main seasons. The seven stars of the Plough remain visible throughout the year anywhere north of latitude 40°N. Even at the latitude (50°N) for which the charts in this book are drawn, many of the stars in the southern portion of the constellation of Ursa Major are hidden below the horizon for part of the year or (particularly in late summer) cannot be seen late in the night.

Polaris and Ursa Minor

The two stars **Dubhe** and **Merak** (α and β Ursae Majoris, respectively), farthest from the 'tail' are known as the 'Pointers'. A line from Merak to Dubhe, extended about five times their separation, leads to the Pole Star, **Polaris**, or α Ursae Minoris. All the stars in the northern sky appear to rotate around it. There are five main stars in the constellation of **Ursa Minor**, and the two farthest from the Pole, **Kochab** and **Pherkad** (β and γ Ursae Minoris, respectively), are known as 'The Guards'.

Cassiopeia

On the opposite of the North Pole from Ursa Major lies **Cassiopeia**. It is highly distinctive, appearing as five stars forming a letter 'W' or 'M' depending on its orientation. Provided the sky is reasonably clear of clouds, you will nearly always be able to see either Ursa Major or Cassiopeia, and thus be able to orientate yourself on the sky.

To find Cassiopeia, start with **Alioth** (ε Ursae Majoris), the first star in the tail of the Great Bear. A line from this star extended through Polaris points directly towards γ Cassiopeiae, the central star of the five.

Cepheus

Although the constellation of **Cepheus** is fully circumpolar, it is not nearly as well-known as Ursa Major, Ursa Minor or Cassiopeia, partly because its stars are fainter. Its shape is rather like the gable-end of a house. The line from the Pointers through Polaris, if extended, leads to **Errai** (γ Cephei) at the 'top' of the 'gable'. The brightest star, **Alderamin** (α Cephei) lies in the Milky Way region, at the 'bottom right-hand corner' of the figure.

Draco

The constellation of **Draco** consists of a quadrilateral of stars, known as the 'Head of Draco' (and also the 'Lozenge'), and a long chain of stars forming the neck and body of the dragon. To find the Head of Draco, locate the two stars **Phecda** and **Megrez** (γ and δ Ursae Majoris) in the Plough, opposite the Pointers.

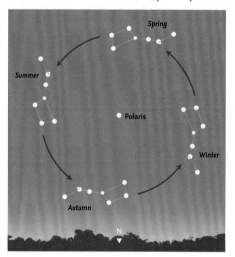

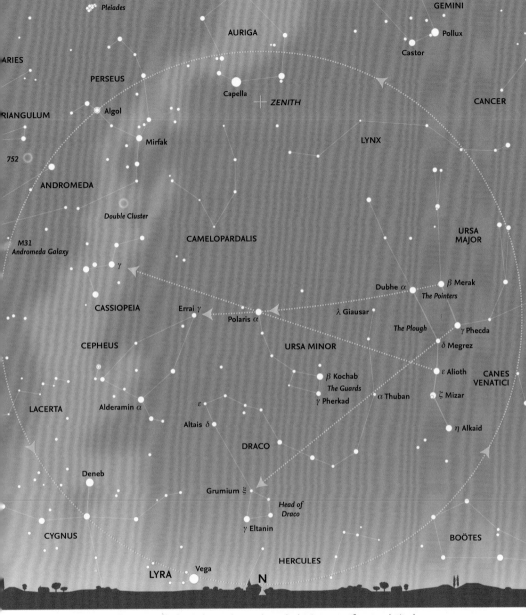

The stars and constellations inside the circle are always above the horizon, seen from our latitude.

Extend a line from Phecda through Megrez by about eight times their separation, right across the sky below the Guards in Ursa Minor, ending at **Grumium** (ξ Draconis) at one corner of the quadrilateral. The brightest star, **Eltanin** (γ Draconis) lies farther to the south. From the head of Draco, the constellation first runs northeast to **Altais** (δ Draconis) and ε Draconis, then doubles back southwards before winding its way through **Thuban** (α Draconis) before ending at **Giausar** (λ Draconis) between the Pointers and Polaris.

The Winter Constellations

The winter sky is dominated by several bright stars and distinctive constellations. The most conspicuous constellation is **Orion**, the main body of which has an hour-glass shape. It straddles the celestial equator and is thus visible from anywhere in the world. The three stars that form the 'Belt' of Orion point down towards the southeast and to **Sirius** (α Canis Majoris), the brightest star in the sky. **Mintaka** (δ Orionis), the star at the northeastern end of the Belt, farthest from Sirius, actually lies just slightly south of the celestial equator.

A line from **Bellatrix** (γ Orionis) at the 'top right-hand corner' of Orion, through **Aldebaran** (α Tauri), past the 'V' of the Hyades cluster, points to the distinctive cluster of bright blue stars known as the **Pleiades**, or the 'Seven Sisters'. Aldebaran is one of the five bright stars that may sometimes be occulted (hidden) by the Moon. Another line from Bellatrix, through **Betelgeuse** (α Orionis), if carried right across the sky, points to the constellation of **Leo**, a prominent constellation in the spring sky.

Six bright stars in six different constellations: **Capella** (α Aurigae), **Aldebaran** (α Tauri), **Rigel** (β Orionis), **Sirius** (α Canis Majoris), **Procyon** (α Canis Minoris) and **Pollux** (β Gemini) form what is sometimes known as the 'Winter Hexagon'. Pollux is accompanied to the northwest by the slightly fainter star of **Castor** (α Gemini), the second 'Twin'.

In a counterpart to the famous 'Summer Triangle', an almost perfect equilateral triangle, the 'Winter Triangle', is formed by Betelgeuse (α Orionis), Sirius (α Canis Majoris) and Procyon (α Canis Minoris).

Several of the stars in this region of the sky show distinctive tints: Betelgeuse (α Orionis) is reddish, Aldebaran (α Tauri) is orange and Rigel (β Orionis) is blue-white.

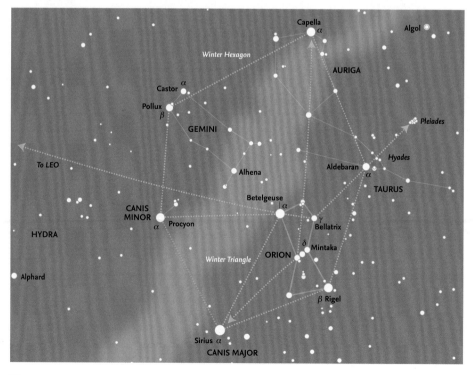

The Spring Constellations

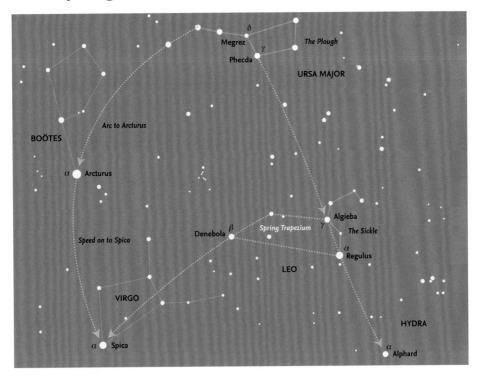

The most prominent constellation in the spring sky is the zodiacal constellation of **Leo**, and its brightest star, **Regulus** (α Leonis), which may be found by extending a line from Megrez and Phecda (δ and γ Ursae Majoris, respectively) – the two stars on the opposite side of the bowl of the Plough from the Pointers – down to the southeast. Regulus forms the 'dot' of the 'backward question mark' known as 'the Sickle'. Regulus, like Aldebaran in Taurus is one of the bright stars that lie close to the ecliptic, and that are occasionally occulted by the Moon. The same line from Ursa Major to Regulus, if continued, leads to **Alphard** (α Hydrae), the brightest star in **Hydra**, the largest of the 88 constellations.

The shape formed by the body of Leo is sometimes known as the 'Spring Trapezium'. At the other end of the constellation from Regulus is **Denebola** (β Leonis), and the line forming the back of the constellation through Denebola points to the bright star **Spica** (α Virginis) in the constellation of **Virgo**. A saying that helps to locate Spica is well-known to astronomers: 'Arc to Arcturus and then speed on to Spica.' This suggests following the arc of the tail of Ursa Major to Arcturus and then on to Spica. **Arcturus** (α Boötis) is actually the brightest star in the northern hemisphere of the sky. (Although other stars, such as Sirius, are brighter, they are all in the southern hemisphere.) Overall, the constellation of **Boötes** is sometimes described as 'kite-shaped' or 'shaped like the letter P'.

Although Spica is the brightest star in the zodiacal constellation of Virgo, the rest of the constellation is not particularly distinct, consisting of a rough quadrilateral of moderately bright stars and some fainter lines of stars extending outwards.

The Summer Constellations

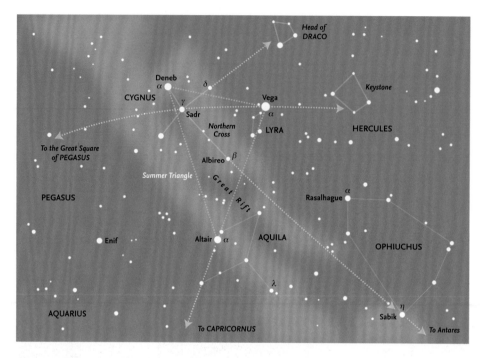

On summer nights, the three bright stars **Deneb** (α Cygni), **Vega** (α Lyrae) and **Altair** (α Aquilae) form the striking 'Summer Triangle'. The constellations of **Cygnus** (the Swan) and **Aquila** (the Eagle) represent birds 'flying' down the length of the Milky Way. This part of the Milky Way contains the **Great Rift**, an elongated dark region, where the light from distant stars is obscured by intervening dust. The dark Rift is clearly visible even to the naked eye.

The most prominent stars of Cygnus are sometimes known as the 'Northern Cross' (as a counterpart to the 'Southern Cross' – the constellation of Crux – in the southern hemisphere). The central line of Cygnus through **Albireo** (β Cygni), extended well to the southwest, points to **Sabik** (η Ophiuchi) in the large, sprawling constellation of **Ophiuchus** (the Serpent Bearer) and beyond to **Antares** (α Scorpii) in the constellation of **Scorpius**. Like Cepheus, the shape of Ophiuchus somewhat resembles the gable-end of a house, and the

brightest star **Rasalhague** (α Ophiuchi) is at the 'apex' of the 'gable'.

A line from the central star of Cygnus, **Sadr** (γ Cygni) through δ Cygni, in the northwestern 'wing' points towards the Head of Draco, and is another way of locating that part of the constellation. Another line from Sadr to Vega indicates the central portion, 'The Keystone', of the constellation of **Hercules**. An arc through the same stars, in the opposite direction, points towards the constellation of **Pegasus**, and more specifically to the 'Great Square of Pegasus'.

Aquila is less conspicuous than Cygnus and consists of a diamond shape of stars, representing the body and wings of the eagle, together with a rather faint star, λ Aquilae, marking the 'head'. **Lyra** (the Lyre) mainly consists of Vega (α Lyrae) and a small quadrilateral of stars to its southeast. Continuation of a line from Vega through Altair indicates the zodiacal constellation of **Capricornus.**

The Autumn Constellations

During the autumn season, the most striking feature is the 'Great Square of Pegasus', an almost perfect rectangle on the sky, forming the main body of the constellation of **Pegasus**. However, the star at the northeastern corner, **Alpheratz**, is actually α Andromedae, and part of the adjacent constellation of **Andromeda**. A line from **Scheat** (β Pegasi) at the northeastern corner of the Square, through **Matar** (η Pegasi), points in the general direction of Cygnus. A crooked line of stars leads from **Markab** (α Pegasi) through **Homam** (ζ Pegasi) and **Biham** (θ Pegasi) to **Enif** (ε Pegasi). A line from Markab through the last star in the Square, **Algenib** (γ Pegasi) points in the general direction of the five stars, including **Menkar** (α Ceti) that form the 'tail' of the constellation of **Cetus** (the Whale). A ring of seven stars lying below the southern side of the Great Square is known as 'the Circlet', part of the constellation of **Pisces** (the Fishes).

Extending the line of the western side of the Great Square towards the south leads to the isolated bright star, **Fomalhaut** (α Piscis Austrini), in the Southern Fish. Following the line of the eastern side of the Great Square towards the north leads to Cassiopeia while, in the other direction, it points towards **Diphda** (β Ceti), which is actually the brightest star in Cetus.

Three bright stars leading northeast from Alpheratz form the main body of the constellation of **Andromeda**. Continuation of that line leads towards the constellation of **Perseus** and **Mirfak** (α Persei). Running southwards from Mirfak is a chain of stars, one of which is the famous variable star **Algol** (β Persei). Farther east, an arc of stars leads to the prominent cluster of the **Pleiades**, in the constellation of **Taurus**.

Between Andromeda and Cetus lie the two small constellations of **Triangulum** (the Triangle) and **Aries** (the Ram).

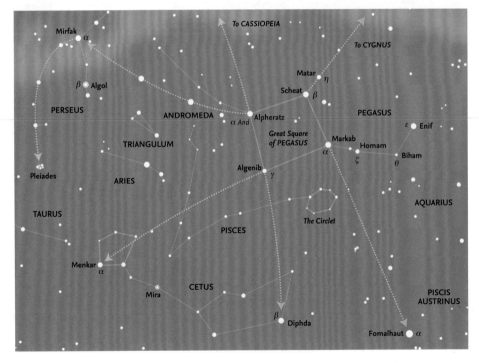

The Moon at First Quarter.

The Moon

The monthly pages include diagrams showing the phase of the Moon for every day of the month, and also indicate the day in the **lunation** (or **age** of the Moon), which begins at New Moon. Although the main features of the surface – the light highlands and the dark maria (seas) – may be seen with the naked eye, far more features may be detected with the use of binoculars or any telescope. The many craters are best seen when they are close to the **terminator** (the boundary between the illuminated and the non-illuminated areas of the surface), when the Sun rises or sets over any particular region of the Moon and the crater walls or central peaks cast strong shadows. Most features become difficult to see at Full Moon, although this is the best time to see the bright ray systems surrounding certain craters. Accompanying the Moon map on the following pages is a list of prominent features, including the days in the lunation when features are normally close to the terminator and thus easiest to see. A few bright features

such as Linné and Proclus, visible when well illuminated, are also listed. One feature, Rupes Recta (the Straight Wall) is readily visible only when it casts a shadow with light from the east, appearing as a light line when illuminated from the opposite direction.

The dates of visibility vary slightly through the effects of **libration**. Because the Moon's orbit is inclined to the Earth's equator and also because it moves in an ellipse, the Moon appears to rock slightly from side to side (and nod up and down). Features near the **limb** (the edge of the Moon) may vary considerably in their location and visibility. (This is easily noticeable with Mare Crisium and the craters Tycho and Plato.) Another effect is that at crescent phases before and after New Moon, the normally non-illuminated portion of the Moon receives a certain amount of light, reflected from the Earth. This **Earthshine** may enable certain bright features (such as Aristarchus, Kepler and Copernicus) to be detected even though they are not illuminated by sunlight.

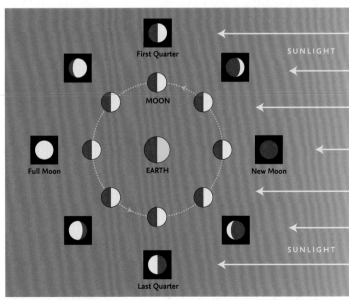

The Moon phases. *During its orbit around the Earth we see different portions of the illuminated side of the Moon's surface.*

Map of the Moon

Abulfeda	6:20
Agrippa	7:21
Albategnius	7:21
Aliacensis	7:21
Alphonsus	8:22
Anaxagoras	9:23
Anaximenes	11:25
Archimedes	8:22
Aristarchus	11:25
Aristillus	7:21
Aristoteles	6:20
Arzachel	8:22
Atlas	4:18
Autolycus	7:21
Barrow	7:21
Billy	12:26
Birt	8:22
Blancanus	9:23
Bullialdus	9:23
Bürg	5:19
Campanus	10:24
Cassini	7:21
Catharina	6:20
Clavius	9:23
Cleomedes	3:17
Copernicus	9:23
Cyrillus	6:20
Delambre	6:20
Deslandres	8:22
Endymion	3:17
Eratosthenes	8:22
Eudoxus	6:20
Fra Mauro	9:23
Fracastorius	5:19
Franklin	4:18
Gassendi	11:25
Geminus	3:17
Goclenius	4:18
Grimaldi	13-14:27-28
Gutenberg	5:19
Hercules	5:19
Herodotus	11:25
Hipparchus	7:21
Hommel	5:19
Humboldt	3:15
Janssen	4:18
Julius Caesar	6:20
Kepler	10:24
Landsberg	10:24
Langrenus	3:17
Letronne	11:25
Linné	6
Longomontanus	9:23

The numbers indicate the age of the Moon when features are usually best visible.

NORTH

Goldschmidt
Philolaus
Barrow
W. Bond
Pythagoras
Carpenter
J. Herschel
60°
MARE FRIGORIS
Harpalus
Plato
SINUS RORIS
Bianchini
Montes Recti
Vallis Alpes
Sharp
Montes Jura
Mons Pico
Montes Alpes
Mairan
SINUS IRIDUM
Mons Rümker
Helicon
Le Verrier
Cassini
Aristillus
MARE IMBRIUM
Autolycus
Delisle
30°
Archimedes
PALUS PUTREDINIS
Vallis Schröteri
Prinz
Timocharis
Struve
Aristarchus
Lambert
Cocon
Herodotus
Euler
Montes Apenninus
Seleucus
Montes Carpatus
Krafft
Mayer
Gay-Lussac
Eratosthenes
MARE VAPORUM
Cardanus
Marius
Copernicus
Stadius
SINUS AESTUUM
Kepler
Reiner
MARE INSULARUM
Bode
Thesnecker
Cavalerius
OCEANUS PROCELLARUM
Reinhold
SINUS MEDII
Hevelius
Landsberg
Gambart
W
60°
30°
Mösting
0°
Flammarion
Grimaldi
Montes Riphaeus
Fra Mauro
Herschel
Sirsalis A
Hansteen
Letronne
MARE COGNITUM
Parry
Bonpland
Ptolemaeus
Billy
Guericke
Davy
Albategnius
Alphonsus
Gassendi
MARE NUBIUM
Alpetragius
Mersenius
Agatharchides
Arzachel
Bullialdus
Thebit
Byrgius
MARE HUMORUM
Birt
Rupes Recta
La Caille
Vieta
Campanus
Purbach
Blanchinus
30°
Vitello
Mercator
Pitatus
Regiomontanus
Werner
PALUS EPIDEMIARUM
Gauricus
Walther
Aliacensis
Cichus
Wurzelbauer
Nonius
Capuanus
Helsius
Orontius
Stöfler
Mee
Wilhelm
Schickard
Tycho
Saussure
Nasireddin
Wargentin
Maginus
Licetus
Phocylides
Schiller
Longomontanus
Clavius
60°
Scheiner
Blancanus
Curtius
Moretus

SOUTH

18

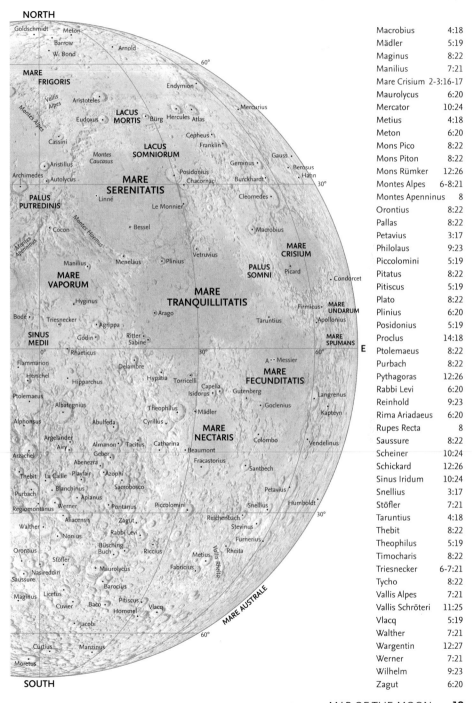

Macrobius 4:18
Mädler 5:19
Maginus 8:22
Manilius 7:21
Mare Crisium 2-3:16-17
Maurolycus 6:20
Mercator 10:24
Metius 4:18
Meton 6:20
Mons Pico 8:22
Mons Piton 8:22
Mons Rümker 12:26
Montes Alpes 6-8:21
Montes Apenninus 8
Orontius 8:22
Pallas 8:22
Petavius 3:17
Philolaus 9:23
Piccolomini 5:19
Pitatus 8:22
Pitiscus 5:19
Plato 8:22
Plinius 6:20
Posidonius 5:19
Proclus 14:18
Ptolemaeus 8:22
Purbach 8:22
Pythagoras 12:26
Rabbi Levi 6:20
Reinhold 9:23
Rima Ariadaeus 6:20
Rupes Recta 8
Saussure 8:22
Scheiner 10:24
Schickard 12:26
Sinus Iridum 10:24
Snellius 3:17
Stöfler 7:21
Taruntius 4:18
Thebit 8:22
Theophilus 5:19
Timocharis 8:22
Triesnecker 6-7:21
Tycho 8:22
Vallis Alpes 7:21
Vallis Schröteri 11:25
Vlacq 5:19
Walther 7:21
Wargentin 12:27
Werner 7:21
Wilhelm 9:23
Zagut 6:20

Eclipses in 2023

Lunar eclipses

There are two lunar eclipses in 2023. Neither is particularly striking. The first, on May 5, is a penumbral eclipse, so will not be readily seen by the naked eye. This eclipse will occur over the Pacific, eastern Asia and Africa. The second, on October 28, is a very small partial eclipse. Only east Africa and parts of the Indian Ocean will see just a very tiny portion of the Moon enter the Earth's umbra.

Solar eclipses

There are also two solar eclipses in 2023. The eclipse of April 20 is known as a hybrid eclipse. The path runs from the Indian Ocean, across Indonesia, and out into the Pacific. A total eclipse (mid-eclipse 04:16) occurs southeast of the island of Flores. Because of the curvature of the Earth, the distance to the Moon is slightly greater over most of the path, where the eclipse will be annular. The second eclipse, on October 14 is annular, and the path runs across the United States, beginning in the northwest, then across Central America (greatest eclipse at 17:59 UT is off the coast of Nicaragua) and then northern South America (Colombia and northern Brazil).

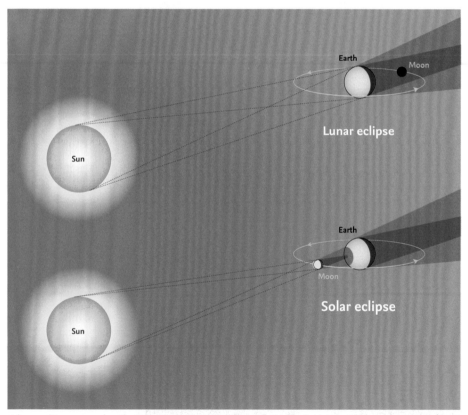

A lunar eclipse occurs when the Moon passes through the Earth's shadow (top). When it passes between the Earth and the Sun (bottom) there is a solar eclipse.

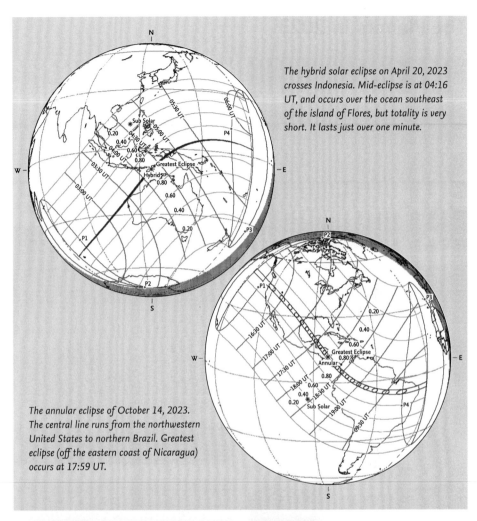

The hybrid solar eclipse on April 20, 2023 crosses Indonesia. Mid-eclipse is at 04:16 UT, and occurs over the ocean southeast of the island of Flores, but totality is very short. It lasts just over one minute.

The annular eclipse of October 14, 2023. The central line runs from the northwestern United States to northern Brazil. Greatest eclipse (off the eastern coast of Nicaragua) occurs at 17:59 UT.

An annular eclipse of the Sun, photographed from California, towards the end of the eclipse at sunset over the Pacific.

The Planets in 2023

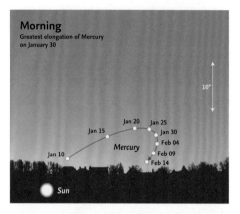

Morning
Greatest elongation of Mercury on January 30

Jan 10
Jan 15
Jan 20
Jan 25
Jan 30
Feb 04
Feb 09
Feb 14
Mercury
10°
Sun

Mercury and Venus

Although **Mercury** comes to greatest elongation on six occasions in 2023, on only four is it reasonably visible above the horizon: January 30, April 11, September 22 and December 4. On two of these (April 11 and September 22), Venus is nearby. **Venus** itself comes to elongation on two occasions: June 4 and October 23. All these events are shown on the accompanying diagrams.

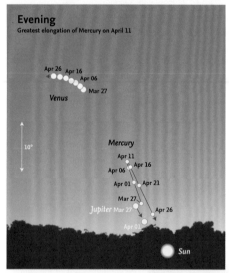

Evening
Greatest elongation of Mercury on April 11

Apr 26
Apr 16
Apr 06
Mar 27
Venus
10°
Mercury
Apr 11
Apr 16
Apr 06
Apr 01
Apr 21
Mar 27
Apr 26
Jupiter Mar 27
Apr 01
Sun

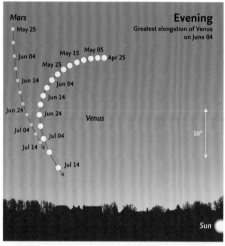

Mars
May 25
Jun 04
May 15
May 05
May 25
Apr 25
Jun 14
Jun 04
Jun 14
Jun 24
Jun 24
Jul 04
Jul 04
Jul 14
Jul 14
Venus
Evening
Greatest elongation of Venus on June 04
10°
Sun

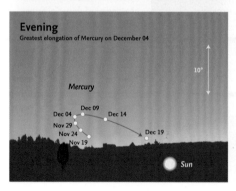

Evening
Greatest elongation of Mercury on December 04

10°
Mercury
Dec 09
Dec 04
Dec 14
Nov 29
Nov 24
Nov 19
Dec 19
Sun

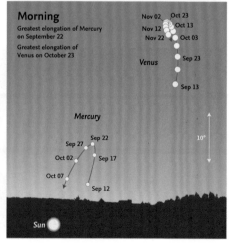

Morning
Greatest elongation of Mercury on September 22
Greatest elongation of Venus on October 23

Nov 02
Oct 23
Nov 12
Oct 13
Nov 22
Oct 03
Sep 23
Venus
Sep 13
Mercury
Sep 27
Sep 22
Oct 02
Sep 17
Oct 07
Sep 12
10°
Sun

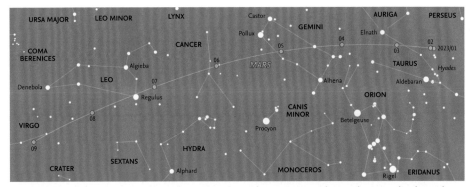

The path of Mars from January to September 2023. Later in the year it is too close to the Sun to be observed.

Mars

Because its orbit lies outside that of the Earth, so taking longer to complete an orbit, **Mars** does not come to opposition every year. Instead it often remains in the sky for many months at a time, slowly moving along the ecliptic. This is the case in 2023, when there is no opposition, but the planet is visible from January to September, as shown on the large chart here. It begins the year in **Taurus** at mag. -1.2. It passes right across the sky, slowly fading to mag. 1.8 in July and August, when it is in **Leo**, and ending the year at mag. 1.4, when it is too close to the Sun to be visible.

Oppositions occur during a period of retrograde motion, when the planet appears to move westwards against the pattern of distant stars. There is no opposition in 2024. Mars actually begins to retrograde on 14 December 2024, and comes to its next opposition on 16 January 2025.

Because of its eccentric orbit, which carries it at very differing distances from the Sun (and Earth), not all oppositions of Mars are equally favorable for observation. The relative positions of Mars and the Earth are shown here. It will be seen that the opposition of 2018 was very close and thus favorable for observation, and that of 2020 was also reasonably good. By comparison, opposition in 2027 will be at a far greater distance, so the planet will appear much smaller.

This image of Mars was obtained by Damian Peach with a 14-inch Celestron telescope on 30 September 2020 when Mars was magnitude -2.5. South is at the top.

The oppositions of Mars between 2018 and 2033. As the illustration shows, there is no opposition in 2023.

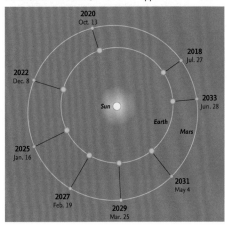

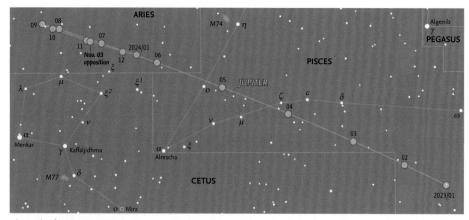

The path of Jupiter in 2023. Jupiter comes to opposition on November 3. Background stars are shown down to magnitude 6.5.

Jupiter and Saturn

In 2023, **Jupiter** spends the early part of the year in **Pisces**, starting in the west, then clips the top of **Cetus** and moves into **Aries** in mid-May. It begins retrograde motion in early September, and comes to opposition on November 3 at mag. -2.9. It continues retrograding to the end of the year, when it is still in Aries. **Saturn** starts the year in **Capricornus**, moving into **Aquarius** in February. It begins its retrograde motion in early July, and comes to opposition on August 27 at mag. 0.4. It continues retrograde motion until late October, reverting to direct motion and finishing the year in the western side of Aquarius.

Jupiter's four large satellites are readily visible in binoculars. Not all four are visible all the time, sometimes hidden behind the planet or invisible in front of it. **Io**, the closest to Jupiter, orbits in just under 1.8 days, and **Callisto**, the farthest away, takes about 16.7 days. In between are **Europa** (c. 3.6 days) and **Ganymede**, the largest, (c. 7.1 days). The diagram (below) shows the satellites' motions around the time of opposition.

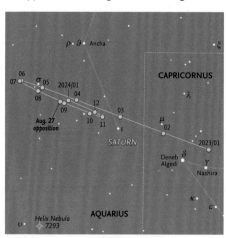

The path of Saturn in 2023. Saturn comes to opposition on August 27. Background stars are shown down to magnitude 6.5.

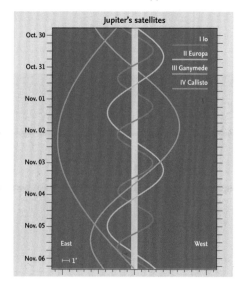

Uranus and Neptune

Uranus is in **Aries** for the whole of 2023. Initially retrograding, it reverts to direct motion in late January, then begins retrograde motion in early September. It reaches opposition (at mag. 5.6) on November 13 (at New Moon, so observation should be feasible). It is still retrograding at the end of the year, when it is mag. 5.6.

Neptune begins the year in **Aquarius** but moves into **Pisces** in March. It begins retrograde motion in mid-July and comes to opposition at mag. 7.8 on September 19. It moves back into Aquarius at the beginning of December, and resumes direct motion in late December when it is mag. 7.9.

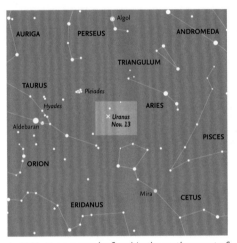

In 2023, Uranus may be found in the southern part of the constellation of Aries.

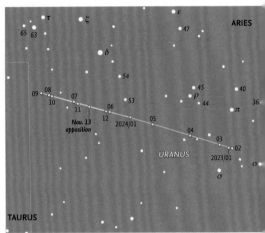

The path of Uranus in 2023. Uranus comes to opposition on November 13. All stars brighter than magnitude 7.5 are shown.

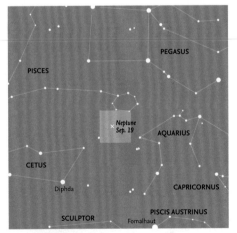

In 2023, Neptune is to be found near the boundary between Aquarius and Pisces. It comes to opposition on September 19.

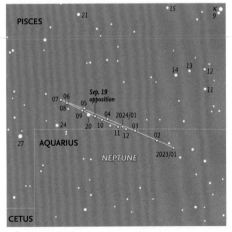

The path of Neptune in 2023. In March it moves across the boundary between the constellations of Aquarius and Pisces. All stars down to magnitude 8.5 are shown.

Minor Planets in 2023

Several minor planets rise above magnitude 9 in 2023, and four come to opposition above mag. 8.5. These are *(2) Pallas* in **Canis Major** on January 8 when it is mag. 7.7; *(1) Ceres* in **Coma Berenices** on March 21 at mag. 6.9; *(8) Flora* in **Aquarius** on August 27 (mag. 8.3) and *(4) Vesta* in the top of **Orion** at mag. 6.4 on December 21. Both (4) Vesta and the dwarf planet (1) Ceres are bright (above mag. 9.0) the whole year.

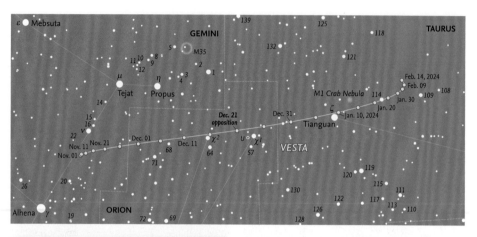

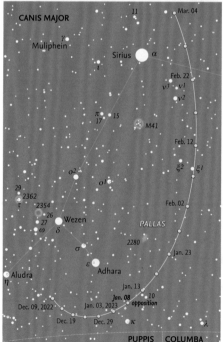

(Top) *The path of the minor planet (4) Vesta around its opposition on December 21 (mag. 6.4). The map also shows the path up to 14 February 2024.*

(Left) *The path of the minor planet (2) Pallas around its opposition on January 8 (mag. 7.7). The map also shows the path for December 2022.*

Both maps show background stars down to magnitude 8.0.

The four small maps at the bottom of these pages show the areas covered by the larger maps in a lighter blue. A white cross marks the position of the minor planet at the day of its opposition.

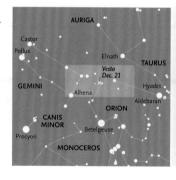

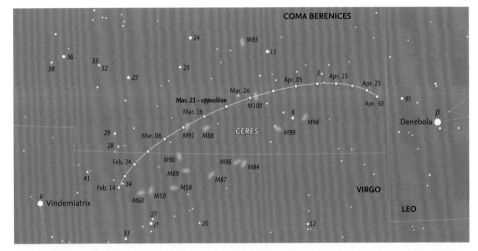

The path of the minor planet (1) Ceres around its opposition on March 21 (mag. 6.9). Background stars are shown down to magnitude 8.0.

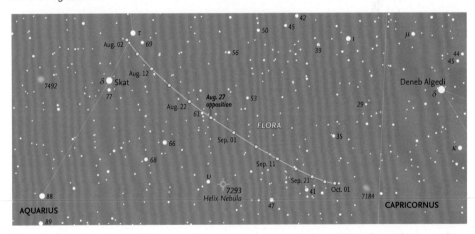

The path of the minor planet (8) Flora around its opposition on August 27 (mag. 8.3). Background stars are shown down to magnitude 9.0.

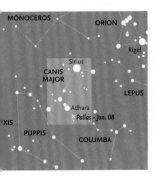

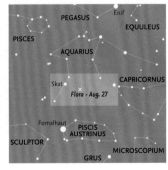

Comets in 2023

Although comets may occasionally become very striking objects in the sky, as was the case with Comet C/2020 F3 NEOWISE in 2020, their occurrence and particularly the existence or length of any tail and their overall magnitude are notoriously difficult to predict. Naturally, it is only possible to predict the return of periodic comets (whose names have the prefix 'P'). Many comets appear unexpectedly (these have names with the prefix 'C'). Bright, readily visible comets such as C/1995 Y1 Hyakutake, C/1995 O1 Hale-Bopp, C/2006 P1 McNaught or C/2020 F3 NEOWISE are rare. Most periodic comets are faint and only a very small number ever become bright enough to be easily visible with the naked eye or with binoculars.

The comets most likely to become visible in late 2023 are 103P/Hartley-2 (August to December) and 62P/Tsuchinshan-1 (November and December). Comet 62P/Tsuchinshan will probably continue bright into early 2024. Comet 2P/Encke will be very bright between October and December, but too close to the Sun to be readily visible.

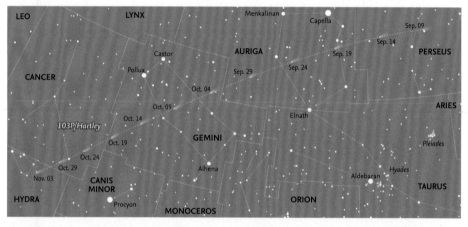

The path of Comet 103P/Hartley from 9 September to 3 November 2023. All stars down to magnitude 6.0 are shown.

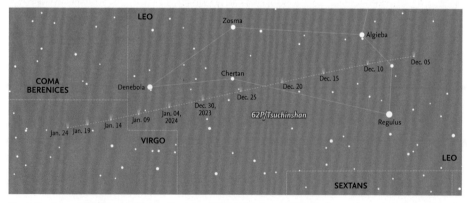

The path of comet 62P/Tsuchinshan in December 2023 and January 2024. Background stars are shown down to magnitude 6.5.

The Comet C/2020 F3 NEOWISE photographed by Nick James on 17 July 2020, showing the light-coloured dust tail, together with the blue ion tail.

Introduction to the Month-by-Month Guide

The monthly charts

The pages devoted to each month contain a pair of charts showing the appearance of the night sky, looking north and looking south. The charts (as with all the charts in this book) are drawn for the latitude of 50°N, so observers farther north will see slightly more of the sky on the northern horizon, and slightly less on the southern. These areas are, of course, those most likely to be affected by poor observing conditions caused by haze, mist or smoke. In addition, stars close to the horizon are always dimmed by atmospheric absorption, so sometimes the faintest stars marked on the charts may not be visible.

The three times shown for each chart require a little explanation. The charts are drawn to show the appearance at 23:00 GMT for the 1st of each month. The same appearance will apply an hour earlier (22:00 GMT) on the 15th, and yet another hour earlier (21:00 GMT) at the end of the month (shown as the 1st of the following month). GMT is identical to the Universal Time (UT) used by astronomers around the world. In Europe, Summer Time is introduced in March, so the March charts apply to 23:00 GMT on March 1, 22:00 GMT on March 15, but 22:00 BST (British Summer Time) on April 1. The change back from Summer Time (in Europe) occurs in October, so the charts for that month apply to 00:00 BST for October 1, 23:00 BST for October 15, and 21:00 GMT for November 1.

The charts may be used for earlier or later times during the night. To observe two hours earlier, use the charts for the preceding month; for two hours later, the charts for the next month.

Meteors

Details of specific meteor showers are given in the months when they come to maximum, regardless of whether they begin or end in other months. Note that not all the respective radiants are marked on the charts for that

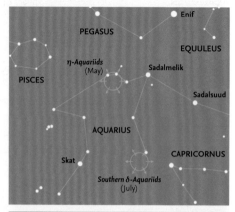

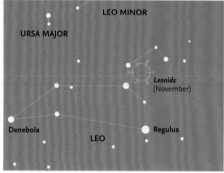

particular month, because the radiants may be below the horizon, or lie in constellations that are not readily visible during the month of maximum. For this reason, special charts for the Eta (η) and Delta (δ) Aquariids (May and July, respectively) and the Leonids (November) are given here. As explained earlier, however, meteors from such showers may still be seen, because the most effective region for seeing meteors is some 40–45° away from the radiant, and that area of sky may well be above the horizon. A table of the best meteor showers visible during the year is also given here. The rates given are based on the properties of the meteor streams, and are those that an experienced observer might see under ideal conditions. Generally, the observed rates will be far less.

Shower	Dates of activity 2023	Date of maximum 2023	Possible hourly rate
Quadrantids	December 28 to January 12	January 3–4	110
April Lyrids	April 14–30	April 22–23	18
η-Aquariids	April 19 to May 28	May 6	50
α-Capricornids	July 3 to August 15	July 30	5
Southern δ-Aquariids	July 12 to August 23	July 30	25
Perseids	July 17 to August 24	August 12–13	100
α-Aurigids	August 28 to September 5	September 1	6
Southern Taurids	September 10 to November 20	October 10–11	5
Orionids	October 2 to November 7	October 21–22	25
Draconids	October 6–10	October 8–9	10
Northern Taurids	October 20 to December 10	November 12–13	5
Leonids	November 6–30	November 17–18	10
Geminids	December 4–20	December 14–15	150
Ursids	December 17–26	December 22–23	10

Meteors that are brighter than magnitude -4 (approximately the maximum magnitude reached by Venus) are known as *fireballs* or *bolides*. Examples are shown on pages 32, 35 and 77. Fireballs sometimes cause sonic booms that may be heard some time after the meteor is seen.

The photographs

As an aid to identification – especially as some people find it difficult to relate charts to the actual stars they see in the sky – one or more photographs of constellations visible in certain specific months are included. It should be noted, however, that because of the limitations of the photographic and printing processes, and the differences between the sensitivity of different individuals to faint starlight (especially in their ability to detect different colours), and the degree to which they have become adapted to the dark, the apparent brightness of stars in the photographs will not necessarily precisely match that seen by any one observer.

The Moon calendar

The Moon calendar is largely self-explanatory. It shows the phase of the Moon for every day of the month, with the exact times (in Universal Time) of New Moon, First Quarter, Full Moon and Last Quarter. Because the times are calculated from the Moon's actual orbital parameters, some of the times shown will, naturally, fall during daylight, but any difference is too small to affect the appearance of the Moon on that date. Also shown is the *age* of the Moon (the day in the *lunation*), beginning at New Moon, which may be used to determine the best time for observation of specific lunar features.

The Moon

The section on the Moon includes details of any lunar or solar eclipses that may occur during the month (visible from anywhere on Earth). Similar information is given about any important occultations. Mainly, however, this section summarizes when the Moon passes close to planets or the five prominent stars close to the ecliptic. The dates when the Moon is closest to the Earth (at *perigee*) and farthest from it (at *apogee*) are shown in the monthly calendars, and only mentioned here when they are particularly significant, such as the nearest and farthest during the year.

The planets and minor planets

Brief details are given of the location, movement and brightness of the planets from Mercury to Saturn throughout the month. None of the planets can, of course, be seen when they are close to the Sun, so such periods are generally noted. All of the planets may sometimes lie on the opposite side of the Sun to the Earth (at superior conjunction), but

A fireball (with flares approximately as bright as the Full Moon), photographed against a weak auroral display by D. Buczynski from Tarbat Ness in Scotland on 22 January 2017.

in the case of the inferior planets, Mercury and Venus, they may also pass between the Earth and the Sun (at inferior conjunction) and are invisible for a period of time, the length of which varies from conjunction to conjunction. Those two planets are normally easiest to see around either eastern or western elongation, in the evening or morning sky, respectively. Not every elongation is favourable, so although every elongation is listed, only those where observing conditions are favourable are shown in the individual diagrams of events.

The dates at which the superior planets reverse their motion (from direct motion to **retrograde**, and retrograde to direct) and of opposition (when a planet generally reaches its maximum brightness) are given. Some planets, especially distant Saturn, may spend most or all of the year in a single constellation. Jupiter and Saturn are normally easiest to see around opposition, which occurs every year. Mars, by contrast, moves relatively rapidly against the background stars and in some years never comes to opposition.

Uranus is normally magnitude 5.7–5.9, and thus at the limit of naked-eye visibility under exceptionally dark skies, but bright enough to be readily visible in binoculars. Because its orbital period is so long (over 84 years), Uranus moves only slowly along the ecliptic, and often remains within a single constellation for a whole year. The chart on page 25 shows its position during 2023.

Similar considerations apply to Neptune, although it is always fainter (generally magnitude 7.8–8.0), still visible in most binoculars. It takes about 164.8 years to complete one orbit of the Sun. As with Uranus, it frequently spends a complete year in one constellation. Its chart is also on page 25.

In any year, few minor planets ever become bright enough to be detectable in binoculars. Just one, (4) Vesta, on rare occasions brightens sufficiently for it to be visible to the naked eye. Our limit for visibility is magnitude 9.0 and details and charts are given for those objects that exceed that magnitude during the year, normally around opposition. To assist in recognition of a planet or minor planet as it moves against the background stars, the latter are shown to a fainter magnitude than the object at opposition. Minor-planet charts for 2023 are on pages 26 and 27.

The ecliptic charts

Although the ecliptic charts are primarily designed to show the positions and motions of the major planets, they also show the motion of the Sun during the month. The light-tinted area shows the area of the sky that is invisible during daylight, but the darker area gives an indication of which constellations are likely to be visible at some time of the night. The closer a planet is to the border between dark and light, the more difficult it will be to see in the twilight.

The monthly calendar

For each month, a calendar shows details of significant events, including when planets are close to one another in the sky, close to the Moon, or close to any one of five bright stars that are spaced along the ecliptic. The times shown are given in Universal Time (UT), always used by astronomers throughout the year, and which is identical to Greenwich Mean Time (GMT). So during the summer months, they do not show Summer Time, which will always be one hour later than the time shown.

The diagrams of interesting events

Each month, a number of diagrams show the appearance of the sky when certain events take place. However, the exact positions of celestial objects and their separations greatly depend on the observer's position on Earth. When the Moon is one of the objects involved, because it is relatively close to Earth, there may be very significant changes from one location to another. Close approaches between planets or between a planet and a star are less affected by changes of location, which may thus be ignored.

The diagrams showing the appearance of the sky are drawn for the latitude of London, so will be approximately correct for most of Britain and Europe. However, for an observer farther north (say, Edinburgh), a planet or star listed as being north of the Moon will appear even farther north, whereas one south of the Moon will appear closer to it – or may even be hidden (occulted) by it. For an observer farther south than London, there will be corresponding changes in the

opposite direction: for a star or planet south of the Moon, the separation will increase, and for one north of the Moon, the separation will decrease. This is particularly important when occultations occur, which may be visible from one location, but not another. However, there are no major occultations visible from Britain at night in 2023.

Ideally, details should be calculated for each individual observer, but this is obviously impractical. In fact, positions and separations are actually calculated for a theoretical observer located at the centre of the Earth.

So the details given regarding the positions of the various bodies should be used as a guide to their location. A similar situation arises with the times that are shown. These are calculated according to certain technical criteria, which need not concern us here. However, they do not necessarily indicate the exact time when two bodies are closest together. Similarly, dates and times are given, even if they fall in daylight, when the objects are likely to be completely invisible. However, such times do give an indication that the objects concerned will be in the same general area of the sky during both the preceding and the following nights.

Data used in this Guide

The data given in this Guide, such as timings and distances between objects, have been computed with a program developed by the US Naval Observatory in Washington DC, widely regarded as the most accurate computation. As such, the data may differ slightly from information given elsewhere.

Key to the symbols used on the monthy star maps.

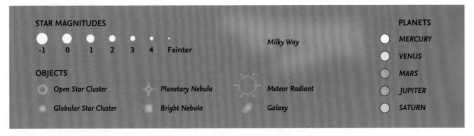

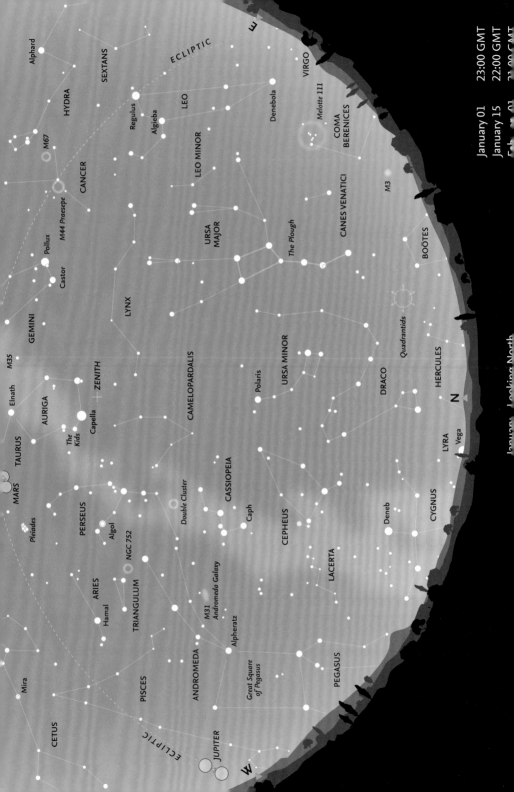

January, Looking North

January 01   23:00 GMT
January 15   22:00 GMT
February 01   21:00 GMT

January – Looking North

A late Quadrantid fireball, photographed from Portmahomack, Ross-shire, Scotland on 15 January 2018 at 23:44 UT.

Ursa Major, the key constellation to navigating the northern sky.

Most of the important circumpolar constellations are easy to see in the northern sky at this time of year. **Ursa Major** stands more-or-less vertically above the horizon in the northeast, with the zodiacal constellation of **Leo** rising in the east. To the north, the stars of **Ursa Minor** lie below **Polaris** (the Pole Star). The head of **Draco** is low on the northern horizon, but may be difficult to see unless observing conditions are good. Both **Cepheus** and **Cassiopeia** are readily visible in the northwest, and even the faint constellation of **Camelopardalis** is high enough in the sky for it to be easily visible.

Near the zenith is the constellation of **Auriga** (the Charioteer), with brilliant **Capella** (α Aurigae), directly overhead. Slightly to the west of Capella lies a small triangle of fainter stars, known as 'The Kids'. (Ancient mythological representations of Auriga show him carrying two young goats.) Together with the northernmost bright star in **Taurus, Elnath** (β Tauri), the body of Auriga forms a large pentagon on the sky, with The Kids lying on the western side. Farther down towards the west are the constellations of **Perseus** and **Andromeda**, and the Great Square of Pegasus is approaching the horizon.

Meteors

The **Quadrantid** meteor shower actually begins in late December (on 28 December 2022) and continues until January 12. It is one of the strongest and most consistent meteor showers. It comes to maximum on the night of January 3 to 4, just before Full Moon, so moonlight will cause some interference. However, the meteors are bright, bluish- or yellowish-white and may reach a maximum rate of 110 per hour. The parent object is minor planet 2003 EH$_1$.

The shower is named after the former constellation **Quadrans Muralis** (the Mural Quadrant), an early form of astronomical instrument. The Quadrantid radiant, marked on the chart, is now within the northernmost part of **Boötes**, roughly halfway between θ Boötis and τ Herculis.

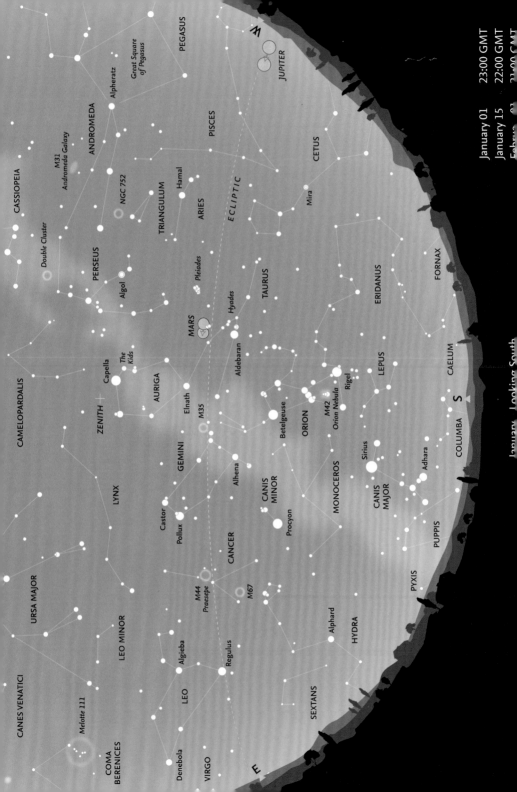

January 01   23:00 GMT
January 15   22:00 GMT
Februar 01   21:00 GMT

January - Looking South

E

S

M

JUPITER

PEGASUS
Great Square
of Pegasus
Alpheratz
ANDROMEDA
CASSIOPEIA
M31
Andromeda Galaxy
NGC 752
TRIANGULUM
Hamal
ARIES
PISCES
CETUS
Double Cluster
PERSEUS
Algol
Pleiades
ECLIPTIC
Mira
MARS
Hyades
Aldebaran
TAURUS
FORNAX
ERIDANUS
CAMELOPARDALIS
The Kids
Capella
ZENITH
AURIGA
Elnath
M35
Betelgeuse
ORION
M42
Orion Nebula
Rigel
LEPUS
CAELUM
COLUMBA
LYNX
GEMINI
Alhena
CANIS
MINOR
MONOCEROS
Sirius
CANIS
MAJOR
Adhara
PUPPIS
Castor
Pollux
Procyon
URSA MAJOR
LEO MINOR
CANCER
M44
Praesepe
M67
PYXIS
HYDRA
Alphard
Melotte 111
Algieba
LEO
Regulus
Denebola
SEXTANS
CANES VENATICI
COMA
BERENICES
VIRGO

January – Looking South

The southern sky is dominated by **Orion**, prominent during the winter months, visible at some time during the night. It is highly distinctive, with a line of three stars that form the 'Belt'. To most observers, the bright star **Betelgeuse** (α Orionis), shows a reddish tinge, in contrast to the brilliant bluish-white **Rigel** (β Orionis). The three stars of the belt lie directly south of the celestial equator. A vertical line of three 'stars' forms the 'Sword' that hangs south of the Belt. With good viewing, the central 'star' appears as a hazy spot, even to the naked eye, and is actually the **Orion Nebula**. Binoculars reveal the four stars of the Trapezium, which illuminate the nebula.

Orion's Belt points up to the northwest towards **Taurus** (the Bull) and orange-tinted **Aldebaran** (α Tauri). Close to Aldebaran, there is a conspicuous 'V' of stars, called the **Hyades** cluster. (Despite appearances, Aldebaran is not part of the cluster.) Farther along, the same line from Orion passes below a bright cluster of stars, the **Pleiades**, or Seven Sisters. Even the smallest pair of binoculars reveals this as a beautiful group of bluish-white stars. The two most conspicuous of the other stars in Taurus lie directly above Orion, and form an elongated triangle with Aldebaran. The northernmost, **Elnath** (β Tauri), was once considered to be part of the constellation of **Auriga**.

The constellation of Orion dominates the sky during this period of the year, and is a useful starting point for recognizing other constellations in the southern sky. Here, orange Betelgeuse, blue-white Rigel and the pinkish Orion Nebula are prominent. Orion can be found in the southern part of the sky (see page 36).

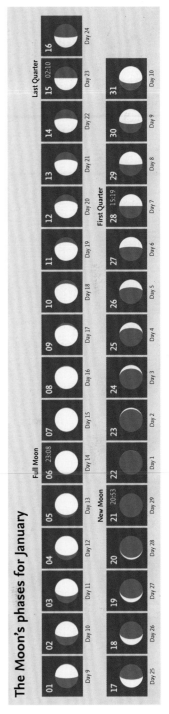

The Moon's phases for January

						Full Moon										Last Quarter
01	02	03	04	05	06 23:08	07	08	09	10	11	12	13	14	15 02:10	16
Day 9	Day 10	Day 11	Day 12	Day 13	Day 14	Day 15	Day 16	Day 17	Day 18	Day 19	Day 20	Day 21	Day 22	Day 23	Day 24

				New Moon								First Quarter			
17	18	19	20	21 20:53	22	23	24	25	26	27	28 15:19	29	30	31	
Day 25	Day 26	Day 27	Day 28	Day 29	Day 1	Day 2	Day 3	Day 4	Day 5	Day 6	Day 7	Day 8	Day 9	Day 10	

January – Moon and Planets

The Earth

The Earth reaches perihelion (the closest point to the Sun in its yearly orbit) on January 4 at 16:17 Universal Time, when its distance is 0.983295578 AU (147,098,855 km).

The Moon

On January 1, the Moon passes 0.7° north of *Uranus* (mag. 5.7). On January 3 it occults *Mars* in *Taurus* (this event is visible only from the Indian Ocean and southern Africa), and the next day is 8.1° north of *Aldebaran*. On January 7, it is 1.9° south of *Pollux* in *Gemini*. On January 10, it passes 4.6° north of *Regulus*, between it and *Algieba*. By January 14, a day before Last Quarter it will be 3.8° north of *Spica* in *Virgo*. As a waning crescent it will be 2.1° north of *Antares* on January 18. On January 20, one day before New Moon, it is 6.9° south of *Mercury* (mag. -0.6) just past inferior conjunction and invisible in the evening sky. On January 23, the thin waxing crescent will be 3.8° south of *Saturn* (mag. 0.8), and a little later, 3.5° south of *Venus*, much brighter at mag. -3.9. On January 31 it occults *Mars* again, this time visible from Central America and the southwestern United States.

The planets

Mercury passes inferior conjunction on January 7. Although it is bright (mag. -3.9), *Venus* is close to the Sun in twilight. *Mars* is in Taurus, slowly fading from mag. -1.2 to -0.3 over the month. It is occulted by the Moon on January 3 and 31. *Jupiter* (average mag. -2.) is in *Pisces* and *Saturn* (mag. 0.8) in *Capricornus*, close to the Sun in evening twilight. *Uranus*, initially retrograding until January 23, is mag. 5.7 in *Aries*. *Neptune* is mag. 7.9 in *Aquarius*. On January 8, the minor planet (2) *Pallas* is at opposition in *Canis Major* at mag. 7.7 (see the chart on page 26). On January 26, minor planet (6) *Hebe* is at opposition in *Hydra* at mag. 8.8.

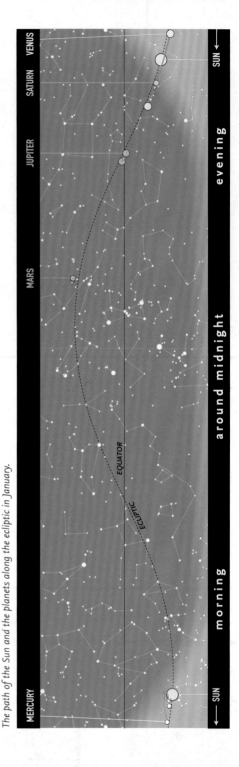

The path of the Sun and the planets along the ecliptic in January.

Calendar for January

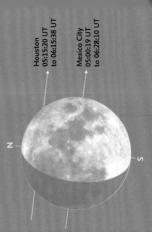

Occultation of Mars

N

S

Houston
05:15:20 UT
to 06:15:38 UT

Mexico City
05:00:19 UT
to 06:28:10 UT

January 31 • *In the early morning Mars is occulted by the Moon. Times of disappearance and reappearance are given for Houston and for Mexico City...*

01–12		Quadrantid meteor shower
01	22:16	Uranus 0.7°S of the Moon
03	19:38	Mars occulted by the Moon
03–04		Quadrantid meteor shower maximum
04	00:53	Aldebaran 8.1°S of the Moon
04	16:17	Earth at perihelion (0.98329578 AU)
06	23:08	Full Moon
07	12:57	Mercury at inferior conjunction
07	14:18	Pollux 1.9°N of the Moon
08	09:19	Moon at apogee = 406,458 km
08	19:02	Minor planet (2) Pallas at opposition (mag. 7.7)
10	12:17	Regulus 4.6°S of the Moon
14	22:37	Spica 3.8°S of the Moon
15	02:10	Last Quarter
18	10:04	Antares 2.1°S of the Moon
20	07:49	Mercury 6.9°N of the Moon
21	20:53	New Moon
21	20:57	Moon at perigee = 356,569 km
22	20:00 *	Saturn (mag. 0.8) 0.4°N of Venus (mag. –3.9)
23	07:21	Saturn 3.8°N of the Moon
23	08:18	Venus 3.5°N of the Moon
25	05:55	Neptune 2.7°N of the Moon
26	02:03	Jupiter 1.8°N of the Moon
26	09:17	Minor planet (6) Hebe at opposition (mag. 8.8)
28	15:19	First Quarter
29	04:09	Uranus 1.0°S of the Moon
30	05:54	Mercury at greatest elongation (25.0°W, mag. –0.1)
31	04:25	Mars occulted by the Moon
31	06:45	Aldebaran 8.3°S of the Moon

These objects are close together for an extended period around this time.

Evening 17:30

Castor
Pollux
Moon

ENE
10°

January 7 • *The nearly Full Moon rises, together with Pollux and Castor.*

Morning 7:00

Sabik
Moon
Antares

SSE
10°

January 18 • *When the Moon rises, it is close to Antares. Sabik (η Oph) is farther southeast.*

Evening 17:15

Venus
Moon
Saturn

SW
10°

January 23 • *The narrow crescent Moon passes Venus (mag. –3.9) and the much fainter Saturn (mag. 1.2).*

Evening 19:00

26
Jupiter
25

SW
W/SW
25°
10°

January 25–26 • *The waxing crescent Moon passes Jupiter in the southwestern sky.*

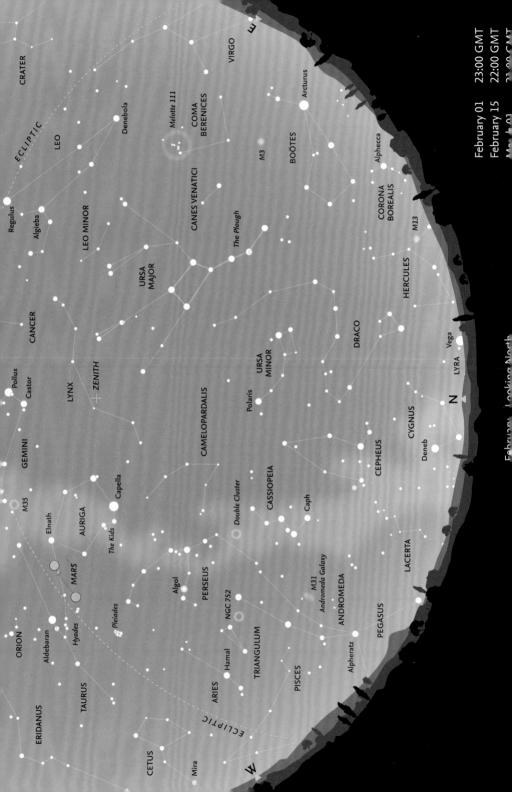

February 01   23:00 GMT
February 15   22:00 GMT
March 01   21:00 GMT

February, Looking North

February – Looking North

The months of January and February are probably the best time for seeing the section of the Milky Way that runs in the northern and western sky from **Cygnus**, low on the northern horizon, through **Cassiopeia**, **Perseus** and **Auriga** and then down through **Gemini** and **Monoceros** (see chart on page 42). Although not as readily visible as the denser star clouds of the summer Milky Way, on a clear night so many stars may be seen that even a distinctive constellation such as Cassiopeia is not immediately obvious.

The head of **Draco** is now higher in the sky and easier to recognize. **Deneb** (α Cygni), the brightest star in **Cygnus**, may just be visible, almost due north at midnight. **Vega** (α Lyrae) in **Lyra** is so low that it is difficult to see, but may become visible later in the night. The constellation of **Boötes** – sometimes described as shaped like a kite, an ice-cream cone, or the letter 'P' – with orange-tinted **Arcturus** (α Boötis), is beginning to clear the eastern horizon. Arcturus, at magnitude -0.05, is the brightest star in the northern hemisphere of the sky. The inconspicuous constellation of **Coma Berenices** is now well above the horizon in the east. The concentration of faint stars at the northwestern corner somewhat resembles a tiny, detached portion of the Milky Way. This is Melotte 111, an open star cluster (which is sometimes called the Coma Cluster, but must not be confused with the important Coma Cluster of galaxies, Abell 1656, mentioned on page 55).

On the other side of the sky, in the northwest, most of the constellation of **Andromeda** is still easily seen, although **Alpheratz** (α Andromedae), the star that forms the northeastern corner of the Great Square of Pegasus – even though it is actually part of Andromeda – is becoming close to the horizon and more difficult to detect. High overhead, at the zenith, try to make out the very faint constellation of

Lynx. It was introduced in 1687 by the famous astronomer Johannes Hevelius to fill the largely blank area between **Auriga**, **Gemini** and **Ursa Major**, and is reputed to be so named because one needed the eyes of a lynx to detect it.

A very large, and frequently ignored, open star cluster, Melotte 111, also known as the Coma Cluster, is readily visible in the eastern sky during February.

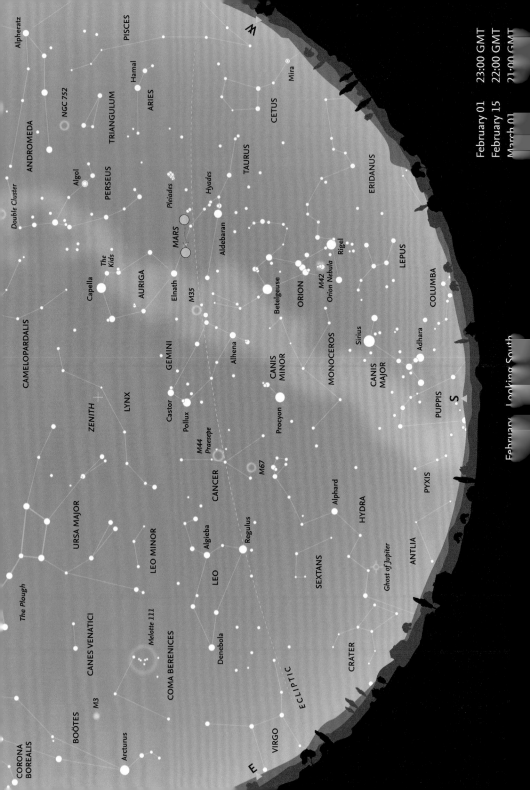

February 01   23:00 GMT
February 15   22:00 GMT
March 01   21:00 GMT

February — Looking South

February – Looking South

Apart from **Orion**, the most prominent constellation visible this month is **Gemini**, with its two lines of stars running southwest towards Orion. Many people have difficulty in remembering which is which of the two stars **Castor** and **Pollux**. Castor (α Geminorum), the fainter star (mag. 1.9), is closer to the North Celestial Pole. Pollux (β Geminorum) is the brighter of the two (mag. 1.2), but is farther away from the Pole. Pollux is one of the first-magnitude stars that may be occulted by the Moon. Castor is remarkable because it is actually a multiple system, consisting of no less than six individual stars.

Orion's belt points down to the southeast towards **Sirius**, the brightest star in the sky (at magnitude -1.4) in the constellation of **Canis Major**, the whole of which is now clear of the southern horizon. Forming an equilateral triangle with **Betelgeuse** in Orion and Sirius in Canis Major is **Procyon**, the brightest star in the small constellation of **Canis Minor**. Between Canis Major and Canis Minor is the faint constellation of **Monoceros**, which actually straddles the Milky Way, which, although faint, has many clusters in this area. Directly east of Procyon is the highly distinctive asterism of six stars that form the

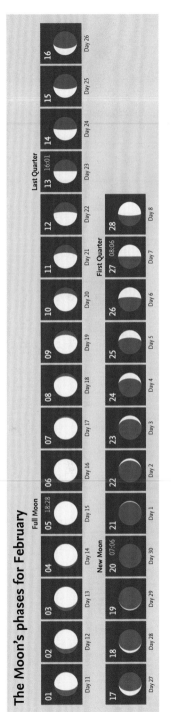

The constellation of **Gemini**: The two brightest stars, visible near the left-hand top corner of the photograph, are Castor (top) and Pollux.

'head' of **Hydra**, the largest of all 88 constellations, and which trails such a long way across the sky that it is only in mid-March around midnight that the whole constellation becomes visible.

The Moon's phases for February

01	02	03	04	05 18:28 Full Moon	06	07	08	09
Day 11	Day 12	Day 13	Day 14 New Moon	Day 15	Day 16	Day 17	Day 18	Day 19
10	11	12	13 16:01 Last Quarter	14	15	16		
Day 20	Day 21	Day 22	Day 23	Day 24	Day 25	Day 26		
17	18	19	20 07:06	21	22	23	24	25
Day 27	Day 28	Day 29	Day 30	Day 1	Day 2	Day 3	Day 4	Day 5
26	27 08:06 First Quarter	28						
Day 6	Day 7	Day 8						

February – Moon and Planets

The Moon

On February 3, two days before Full Moon, the Moon passes 1.9° south of **Pollux** in **Gemini**. One day after Full Moon, it is 4.5° north of **Regulus** in **Leo**. On February 11, it is 3.5° north of **Spica** in **Virgo**. On February 14, one day after Last Quarter, the Moon is 1.8° north of **Antares** in **Scorpius**. On February 19, one day before New Moon, it is 3.7° south of **Saturn**, but this will be lost in twilight. The same problem will apply on February 21, when the Moon is 2.5° south of **Neptune**. The next day, the Moon is 2.1° south of Venus, which is mag. -3.9, so will be visible in the evening twilight. Later the same day, the Moon is 1.2° south of **Jupiter** (mag. -2.1) in **Pisces**. On February 28, the Moon is 1.1° north of Mars, in **Taurus**.

Occultations

In 2023 there are no lunar occultations of the five brightest stars near the ecliptic: **Aldebaran**, **Antares**, **Pollux**, **Regulus** and **Spica**, nor of stars in the **Pleiades**. There are two full occultations of **Mars**, the first on January 3, visible only from the Indian Ocean and southern Africa, and the second, on January 31, visible from Central America and the southwestern United States. The start of an occultation of **Venus** on March 24 may be visible from part of southeastern Asia. There are no major occultations of **Jupiter** and none of **Saturn**.

The planets

Mercury is invisible in twilight. **Venus** is bright (mag. -3.9) and will become visible in the evening sky towards the end of the month. **Mars** fades from mag. -0.3 to mag. 0.4 over the month and is in **Taurus**. **Jupiter** is mag. -2.1 and is in **Pisces**, but **Saturn** is much closer to the Sun, and will be lost in twilight. **Uranus** remains in **Aries** at mag. 5.8, and **Neptune** in **Pisces** at mag. 7.9 to 8.0, close to evening twilight.

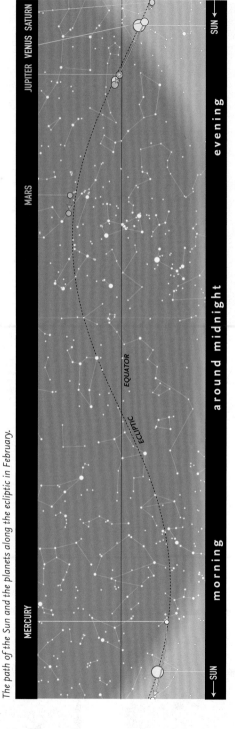

The path of the Sun and the planets along the ecliptic in February.

Calendar for February

* These objects are close together for an extended period of this time.

Evening 21:00

Castor • Pollux • Moon Procyon • SE 60°— 10°

February 3 • The Moon is close to Castor and Pollux, high in the southeast.

Morning 5:00

Moon • Spica SSW 10°

February 11 • The waning gibbous Moon passes Spica, in the south-southwest.

Morning 6:00

14 • Sabik Antares 15 SSE S 10°

February 14–15 • The Moon passes Antares. Sabik (η Oph) is nearby.

Evening 18:00

Algenib • Jupiter • Moon Venus Markab • Diphda • WSW SW 10°

February 22 • The Moon is in the company of Jupiter and Venus, surrounded by Diphda (β Cet), Algenib (γ Peg) and Markab (α Peg).

After midnight 1:30

Capella • Elnath • Mars Moon Alhena • WNW Betelgeuse W 10°

February 28 • The Moon is close to Mars, in the west-northwest. Elnath (β Tau), Alhena (γ Gem), Capella (α Aur) and Betelgeuse (α Ori) are all nearby.

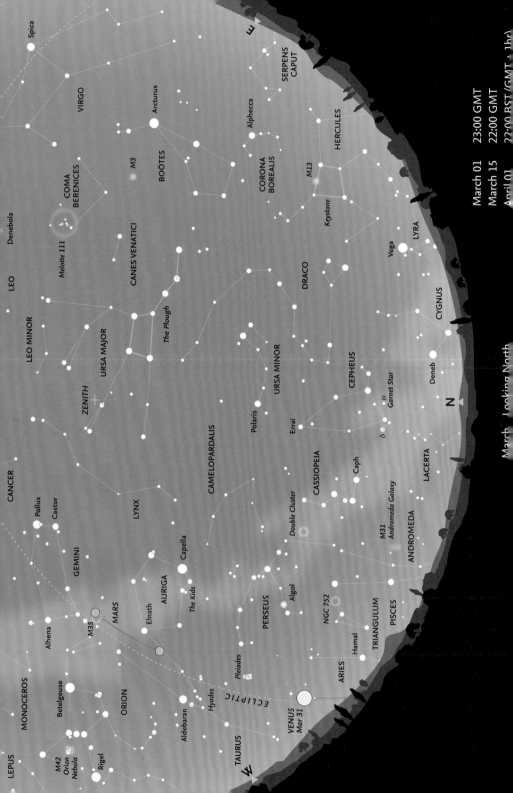

March 01   23:00 GMT
March 15   22:00 GMT
April 01   22:00 BST (GMT + 1hr)

March – Looking North

March – Looking North

The Sun crosses the celestial equator on Monday, March 20, at the vernal equinox, when day and night are of almost equal length, and the northern season of spring is considered to begin. (The hours of daylight and darkness change most rapidly around the equinoxes in March and September.) It is also in March that Daylight Saving Time (DST) begins (on Saturday–Sunday, March 25–26) so the charts show the appearance at 11 p.m. for March 1 and 10 p.m. (BST) for April 1. (In Europe, Daylight Saving Time is introduced on the same date.)

Early in the month, the constellation of **Cepheus** lies almost due north, with the distinctive 'W' of **Cassiopeia** to its west. Cepheus lies across the border of the Milky Way and is often described as like the gable-end of a house or a church tower and steeple. Despite the large number of stars revealed at the base of the constellation by binoculars, one star stands out because of its deep red colour. This is **μ Cephei**, also known as the Garnet Star, because of its striking colour. It is a truly gigantic star, a red supergiant, and one of the largest stars known. It is about 1,400 times the diameter of the Sun, and if placed in the Solar System would extend beyond the orbit of Jupiter. (Betelgeuse, in Orion, is also a red supergiant, but it is 'only' about 500 times the diameter of the Sun.)

Another famous, and very important star in Cepheus is **δ Cephei**, which is the prototype for the class of variable stars known as Cepheids. These giant stars show a regular variation in their luminosity, and there is a direct relationship between the period of the changes in magnitude and the stars' actual luminosity. From a knowledge of the period of any Cepheid, its actual luminosity – known as its absolute magnitude – may be derived. A comparison of its apparent magnitude on the sky and its absolute magnitude enables the star's exact distance to be determined. Once the distances to the first Cepheid variables had been established,

examples in more distant galaxies provided information about the scale of the universe. Cepheid variables are the first major 'rung' in the cosmic distance ladder. Both important stars are shown on the accompanying chart and in the photograph on page 49.

Below Cepheus to the east (to the right), it may be possible to catch a glimpse of **Deneb** (α Cygni), just above the horizon. Slightly farther round towards the northeast, **Vega** (α Lyrae) is marginally higher in the sky. From southern Britain, Deneb is just far enough north not to be circumpolar (although difficult to see in January and February because it is so low on the northern horizon). Vega, by contrast, farther south, is completely hidden during the depths of winter.

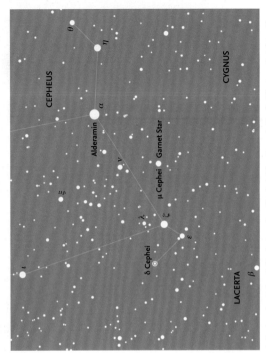

A finder chart for δ Cephei and μ Cephei (the Garnet Star). All stars brighter than magnitude 7.5 are shown. A photograph of the area is on page 49.

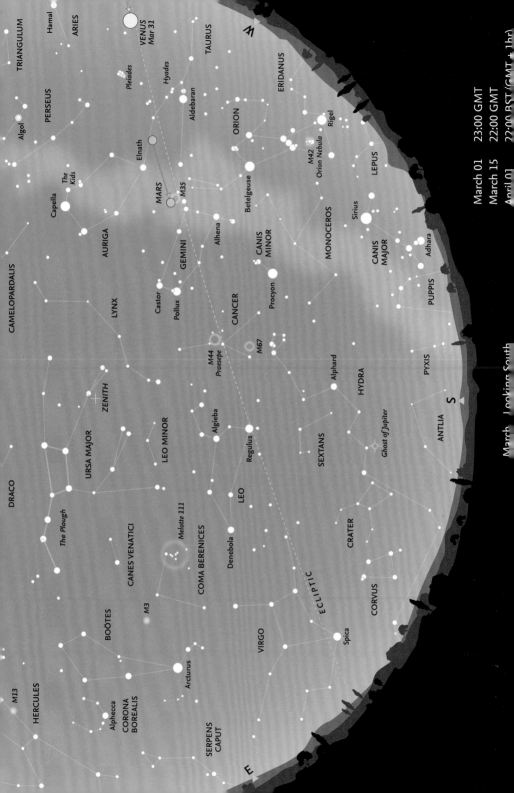

March — Looking South

March 01   23:00 GMT
March 15   22:00 GMT
April 01   22:00 BST (GMT + 1hr)

March – Looking South

Due south at 22:00 at the beginning of the month, lying between the constellations of **Gemini** in the west and **Leo** in the east, and fairly high in the sky above the head of **Hydra**, is the faint, and rather undistinguished zodiacal constellation of **Cancer**. Rather like the triskelion, the symbol for the Isle of Man, it has three 'legs' radiating from the centre, where there is an open cluster, M44 or **Praesepe** ('the Manger' but also known as 'the Beehive'). On a clear night this cluster, known since antiquity, is just a hazy spot to the naked eye, but appears in binoculars as a group of dozens of individual stars.

Also prominent in March is the constellation of Leo, with the 'backward question mark' (or 'Sickle') of bright stars forming the head of the mythological lion. **Regulus** (α Leonis) – the 'dot' of the 'question mark' or the handle of the sickle and the brightest star in Leo – lies very close to the ecliptic and is one of the few first-magnitude stars that may be occulted by the Moon. However, there are no occultations of Regulus in 2023, nor of any of the other four bright stars near the ecliptic.

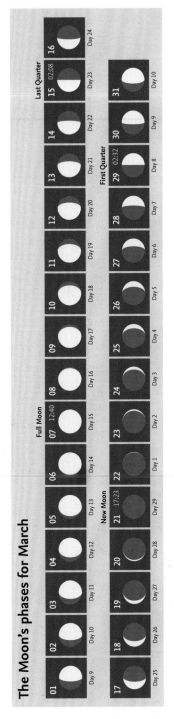

The constellation of Cepheus, with the 'W' of Cassiopeia on the left. The locations of both δ and μ Cephei are shown on the chart on page 47.

The Moon's phases for March

						Full Moon		
01	02	03	04	05	06	07 12:40	08	09
Day 9	Day 10	Day 11	Day 12	Day 13	Day 14	Day 15	Day 16	Day 17

New Moon

10	11	12	13	14	Last Quarter 15 02:08	16
Day 18	Day 19	Day 20	Day 21	Day 22	Day 23	Day 24

17	18	19	20	21 17:23	22	23	24	25	26
Day 25	Day 26	Day 27	Day 28	Day 29	Day 1	Day 2	Day 3	Day 4	Day 5

First Quarter

27	28	29 02:32	30	31
Day 6	Day 7	Day 8	Day 9	Day 10

March – Moon and Planets

The Moon

On March 3, the waxing gibbous Moon is 1.7° south of **Pollux** (mag. 1.1) the brightest star in **Gemini**. On March 6, one day before Full Moon, it is 4.5° north of **Regulus** (mag. 1.4) in **Leo**. By March 10 it is 3.4° north of **Spica** and by March 14, one day before Last Quarter, it is 1.6° north of **Antares** in **Scorpius**. On March 19, the waning crescent is 3.6° south of **Saturn** (mag. 0.8). At New Moon, on March 21, it is 2.4° south of faint **Neptune** (mag. 8.0). The next day in twilight, it passes south of **Mercury**, and then 0.5° south of **Jupiter** (mag. −2.1). On March 24, the Moon occults **Venus**, partly visible from southeast Asia. On March 26, the waxing crescent is 8.7° north of **Aldebaran** in **Taurus**. It passes 2.3° north of **Mars** on March 28 and on March 30 is again 1.6° south of **Pollux.**

The Planets

Mercury is too close to the Sun to be seen. It reaches superior conjunction, on the far side of the Sun, on March 17. **Venus**, in the evening sky, is very bright (mag. −3.9 to −4.0), but too close to the Sun to be easily visible. **Mars** is initially mag. 0.4 in **Taurus**, but moves into **Gemini** and fades to mag. 1.0. **Jupiter** is in **Pisces**, but is too close to the Sun to be readily visible this month. **Saturn** is in **Aquarius** and too deep in the morning twilight to be seen. **Uranus** is in **Aries** at mag. 5.8 and **Neptune** (mag. 8.0) is in **Pisces** and comes to conjunction on March 15.

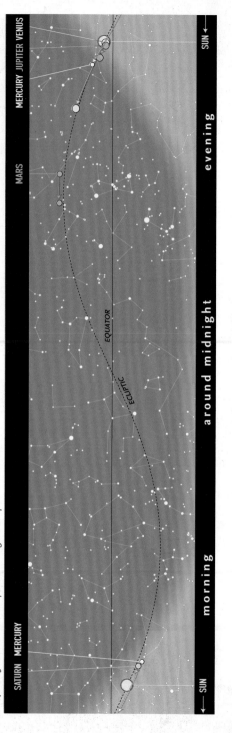

The path of the Sun and the planets along the ecliptic in March.

Calendar for March

2	10:00 *	Saturn (mag. 0.9) 0.9°N of Mercury (mag. −0.6)
2	11:00 *	Jupiter (mag. −2.1) 0.5°S of Venus (mag. −3.9)
3	02:48	Pollux 1.7°N of the Moon
3	15:00	Moon at apogee = 405,889 km
6	00:46	Regulus 4.5°S of the Moon
7	12:40	Full Moon
10	10:45	Spica 3.4°S of the Moon
14	00:56	Antares 1.6°S of the Moon
15	02:08	Last Quarter
15	23:39	Neptune at superior conjunction
16	15:00 *	Neptune (mag. 8.0) 0.4°N of Mercury (mag. −1.8)
17	10:45	Mercury at superior conjunction
19	15:12	Moon at perigee = 362,697 km
19	15:22	Saturn 3.6°N of the Moon
20	21:24	Vernal Equinox
21	06:47	Neptune 2.4°N of the Moon
21	07:41	Dwarf planet Ceres at opposition (mag. 6.9)
21	17:23	New Moon
22	00:10	Mercury 1.8°N of the Moon
22	19:56	Jupiter 0.5°N of the Moon
24	10:26	Venus occulted by the Moon
25	00:39	Uranus 1.5°S of the Moon
26	01:00	European Daylight Saving Time (British Summer Time, BST) begins
26	22:05	Aldebaran 8.7°S of the Moon
28	13:16	Mars 2.3°S of the Moon
28	15:00 *	Mercury (mag. −1.4) 1.5°N of Jupiter (mag. −2.1)
29	02:32	First Quarter
30	10:02	Pollux 1.6°N of the Moon
31	06:00 *	Uranus (mag. 5.8) 1.3°S of Venus (mag. −4.0)
31	11:17	Moon at apogee = 404,919 km

These objects are close together for an extended period...

Evening 18:30

Venus
Jupiter
WSW

March 2 • *After sunset, Venus (mag. −3.9) and Jupiter (mag. −2.1), are close together in the western sky.*

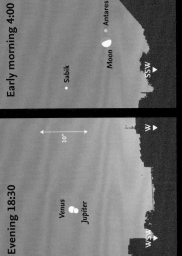

Early morning 4:00

• Sabik
Moon • Antares
SSW

March 14 • *The Moon and Antares are side-by-side in the south-southwest. Sabik is nearby.*

Evening 18:45

• Hamal
24 Venus
Jupiter
22
23
W

March 22–24 • *In the evening twilight the narrow crescent Moon passes Jupiter and Venus.*

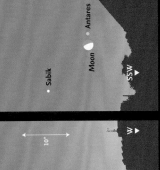

Evening 20:00 (BST)

Capella •
60°
• Elnath
27
28
Mars •
Alhena •
SW
WSW

March 27–28 • *High in the southwest, the Moon passes Elnath and Mars. Capella and Alhena are...*

Morning 3:00 (BST)

Castor •
Pollux •
Moon
Capella •
WNW
NW

March 30 • *In the early morning, the Moon forms a nice triangle with Castor and Pollux. Capella is far...*

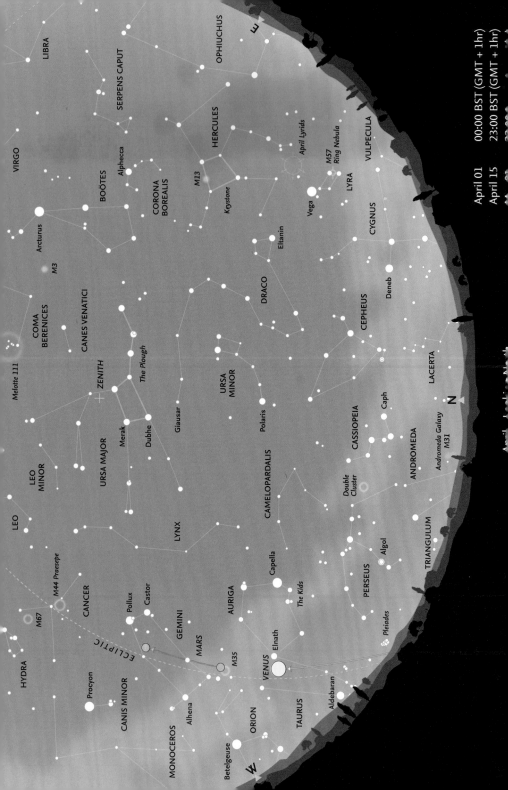

April – Looking North

Cygnus and the brighter regions of the Milky Way are now becoming visible, running more-or-less parallel with the horizon in the early part of the night. Rising in the northeast is the small constellation of **Lyra** and the distinctive 'Keystone' of **Hercules** above it. This asterism is very useful for locating the bright globular cluster M13 (see map on page 59), which lies on one side of the quadrilateral. The winding constellation of **Draco** weaves its way from the four stars that mark its 'head', on the border with Hercules, to end at **Giausar** (λ Draconis) between **Polaris** (α Ursae Minoris) and the 'Pointers', **Dubhe** and **Merak** (α and β Ursae Majoris, respectively). **Ursa Major** is 'upside down' high overhead, near the zenith. The constellation of **Gemini** stands almost vertically in the west. **Auriga** is still clearly seen in the northwest, but, by the end of the month, the southern portion of **Perseus** is starting to dip below the northern horizon. The very faint constellation of **Camelopardalis** lies in the northwest between Polaris and the constellations of Auriga and Perseus.

Meteors

There is one moderate meteor shower this month, the **Lyrids** (often known as the **April Lyrids**). This year, the shower begins on April 14, one day after Last Quarter, and comes to maximum on April 22–23, just after New Moon, so conditions are favourable. There is another, stronger shower, the **η-Aquariids**, that begins on April 19, when the Moon is a day before New Moon, but continues until May 28, with maximum on May 6, at Full Moon.

The constellation of Draco, winding round Ursa Minor, with Polaris center left.

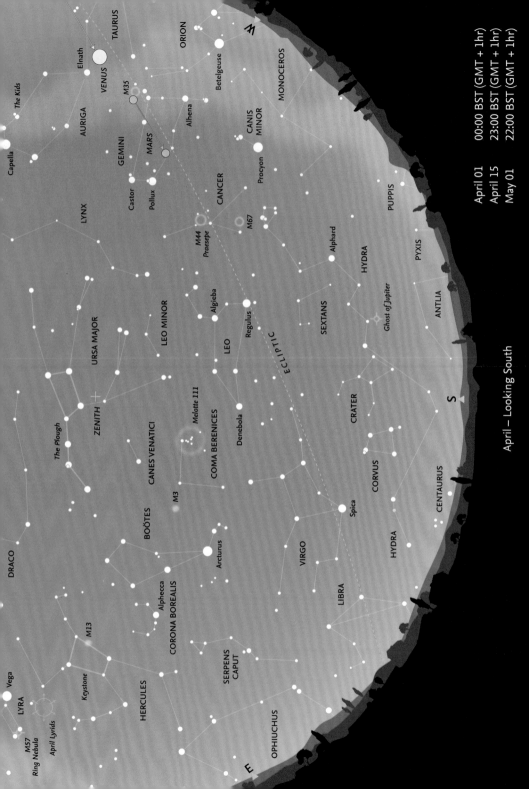

April – Looking South

April 01   00:00 BST (GMT + 1hr)
April 15   23:00 BST (GMT + 1hr)
May 01   22:00 BST (GMT + 1hr)

April – Looking South

Leo is the most prominent constellation in the southern sky in April, and vaguely looks like the creature after which it is named. **Gemini**, with **Castor** and **Pollux**, remains clearly visible in the west, and **Cancer** lies between the two constellations. To the east of Leo, the whole of **Virgo**, with **Spica** (α Virginis) its brightest star, is well clear of the horizon. Below Leo and Virgo, the complete length of **Hydra** is visible, running beneath both constellations, with **Alphard** (α Hydrae) halfway between **Regulus** and the southwestern horizon. Farther east, the two small constellations of **Crater** and the rather brighter **Corvus** lie between Hydra and Virgo.

Boötes and **Arcturus** are prominent in the eastern sky, together with the circlet of **Corona Borealis**, framed by Boötes and the neighbouring constellation of **Hercules**. Between Leo and Boötes lies the constellation of **Coma Berenices**, notable for being the location of the open cluster Melotte 111 (see page 41) and the Coma Cluster of galaxies (Abell 1656). There are about 1000 galaxies in this cluster, which is located near the North Galactic Pole, where we are looking out of the plane of the Galaxy and are thus able to see deep into space. Only about ten of the brightest galaxies in the Coma Cluster are visible with the largest amateur telescopes.

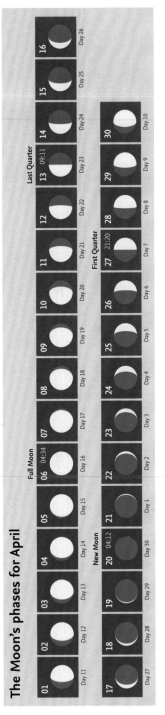

The distinctive constellation of Leo, with Regulus and 'The Sickle' on the west. Algieba (γ Leonis), north of Regulus, appearing double, is a multiple system of four stars.

The Moon's phases for April

					Full Moon				
01	02	03	04	05	06 04:34	07	08	09	10
Day 11	Day 12	Day 13	Day 14	Day 15	Day 16	Day 17	Day 18	Day 19	Day 20

	New Moon		Last Quarter						
11	12	13 09:11	14	15	16				
Day 21	Day 22	Day 23	Day 24	Day 25	Day 26				

	New Moon					First Quarter			
17	18	19	20 04:12	21	22	23	24	25	26
Day 27	Day 28	Day 29	Day 30	Day 1	Day 2	Day 3	Day 4	Day 5	Day 6

	First Quarter			
27 21:20	28	29	30	
Day 7	Day 8	Day 9	Day 10	

April – Moon and Planets

The Moon

On April 2, the waxing gibbous Moon is 4.6° north of **Regulus** (mag. 1.4). On April 6, just after Full Moon, it is 3.3° north of **Spica** in **Virgo**. By April 10, it is 1.5° north of **Antares** in the morning sky. On April 16, it passes 3.5° south of **Saturn**, low in the evening sky. On April 19, one day before New Moon, it is 0.1° north of **Jupiter** in **Pisces**. On April 20, there is a hybrid **eclipse**, visible from Indonesia (see page 20). The next day, the Moon is 1.9° south of **Mercury** and, later, 1.7° north of **Uranus**, both too faint to be readily visible. By April 23 the waxing crescent Moon is 8.8° north of **Aldebaran** in **Taurus**. Later that day it is 1.3° north of brilliant **Venus** (mag. -4.1). On April 26, the Moon is 3.2° north of **Mars** in **Gemini** and then 1.5° south of **Pollux**. By April 29, it is again 4.6° north of **Regulus**, as it was at the beginning of the month.

The planets

Mercury is close to the Sun. It passed superior conjunction on March 17, and comes to greatest eastern elongation (19.5° from the Sun at mag. -0.0) on April 11. **Venus** is very bright (mag. -4.0 to -4.2) in the evening sky. **Mars** (mag. 1.1 to 1.3) moves across **Gemini** over the month. **Jupiter** is lost in the twilight in **Pisces**. **Saturn** (mag. 1.0.) is in **Aquarius**, very low in the morning twilight. **Uranus** (mag. 5.8 to 5.9) is slowly moving eastwards in **Aries**. **Neptune** is still in Pisces, at mag. 8.0 to 7.9.

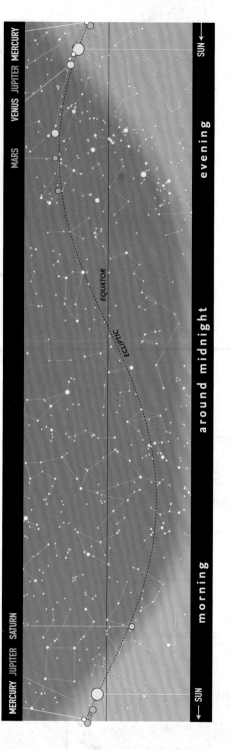

The path of the Sun and the planets along the ecliptic in April.

Calendar for April

02	08:01	Regulus 4.6°S of the Moon
06	04:34	Full Moon
06	17:23	Spica 3.3°S of the Moon
10	06:25	Antares 1.5°S of the Moon
11	22:10	Mercury at greatest elongation (19.5°E, mag. –0.0)
11	22:27	Jupiter at superior conjunction
13	09:11	Last Quarter
14–30		April Lyrid meteor shower
16	02:24	Moon at perigee = 367,968 km
16	03:49	Saturn 3.5°N of the Moon
17	17:24	Neptune 2.3°N of the Moon
19–May 28		η-Aquariid meteor shower
19	17:31	Jupiter 0.1°S of the Moon
20	04:12	New Moon
20	04:36	Hybrid solar eclipse
21	07:05	Mercury 1.9°N of the Moon
21	13:00	Uranus 1.7°S of the Moon
22–23		April Lyrid meteor shower maximum
23	07:17	Aldebaran 8.8°S of the Moon
23	13:03	Venus 1.3°S of the Moon
26	02:18	Mars 3.2°S of the Moon
26	18:03	Pollux 1.5°N of the Moon
27	21:20	First Quarter
28	06:43	Moon at apogee = 404,299 km
29	16:00	Regulus 4.6°S of the Moon

Early morning 4:00 (BST)

April 2–3 • In the early morning, the Moon passes between Regulus and Algieba (γ Leo).

Evening 22:30 (BST)

April 6 • Shortly after Full Moon, it passes Spica, in the southeastern sky.

Morning 5:30 (BST)

April 10 • The Moon is close to Antares. Sabik is about ten degrees higher.

Evening 20:45 (BST)

April 21–23 • In the evening twilight, the narrow crescent Moon passes the Pleiades, Aldebaran and Venus.

Evening 20:45 (BST)

April 29 • The Moon is between Regulus and Algieba, high in the south. Alphard (α Hya) is about twenty degrees closer to the horizon.

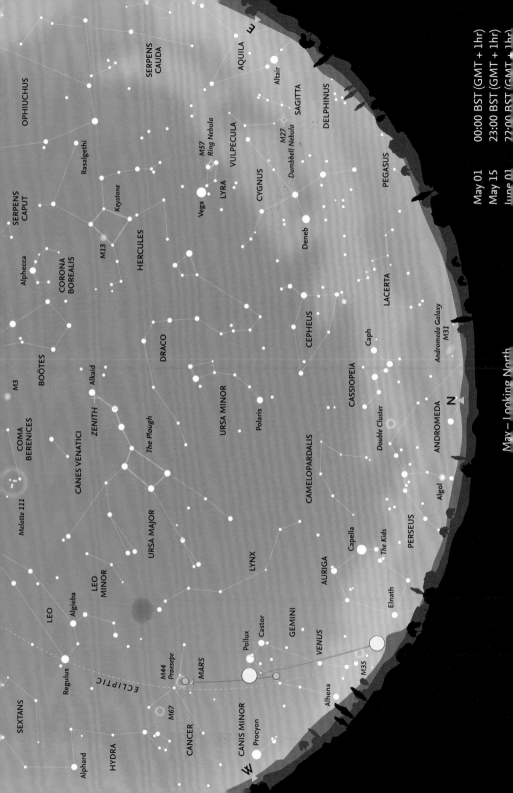

May – Looking North

May – Looking North

Cassiopeia is now low over the northern horizon and, to its west, the southern portions of both **Perseus** and **Auriga** are becoming difficult to observe, although the **Double Cluster**, between Perseus and Cassiopeia is still clearly visible. The **Andromeda Galaxy** is now too low to be readily visible. The constellations of **Lyra**, **Cepheus**, **Ursa Minor** and the whole of **Draco** are well placed in the sky. **Gemini**, with **Castor** and **Pollux**, is sinking towards the western horizon. **Capella** (α Aurigae) and the asterism of **The Kids** are still clear of the horizon.

In the east, two of the stars of the 'Summer Triangle', **Vega** (α Lyrae) and **Deneb** (α Cygni), are clearly visible, and the third star, **Altair** in **Aquila**, is beginning to climb above the horizon. The whole of **Cygnus** is now visible. The sprawling constellation of **Hercules** is high in the east and the brightest globular cluster in the northern hemisphere, M13, is visible to the naked eye on the western side of the asterism known as the **Keystone**.

Three faint constellations may be identified before the lighter nights of summer make them difficult objects. Below Cepheus, in the northeastern sky is the zig-zag constellation of **Lacerta**, while to the west, above Perseus and Auriga is **Camelopardalis** and, farther west, the line of faint stars forming **Lynx**.

Later in the night (and in the month) the westernmost stars of **Pegasus** begin to come into view, while the stars of **Andromeda** are skimming the northeastern horizon and the Andromeda Galaxy may become visible. High overhead, **Alkaid** (η Ursae Majoris), the last star in the 'tail' of the Great Bear, is close to the zenith, while the main body of the constellation has swung round into the western sky.

Meteors

The **η-Aquariids** are one of the two meteor showers associated with Comet 1P/Halley (the other being the Orionids, in October). The η-Aquariids are not particularly favourably placed for northern-

hemisphere observers, because the radiant is near the celestial equator, near the 'Water Jar' in Aquarius, well below the horizon until late in the night (around dawn). However, meteors may still be seen in the eastern sky even when the radiant is below the horizon. There is a radiant map for the η-Aquariids on page 30.

Their maximum in 2023, on May 6, occurs just after Full Moon, so is particularly unfavourable. Maximum hourly rate is about 50–55 per hour and a large proportion (about 25 per cent) of the meteors leave persistent trains.

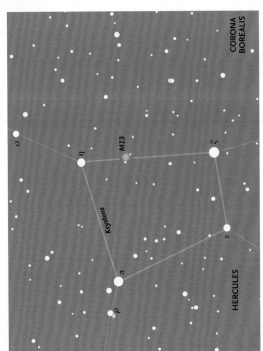

A finder chart for M13, the finest globular cluster in the northern sky. All stars down to magnitude 7.5 are shown.

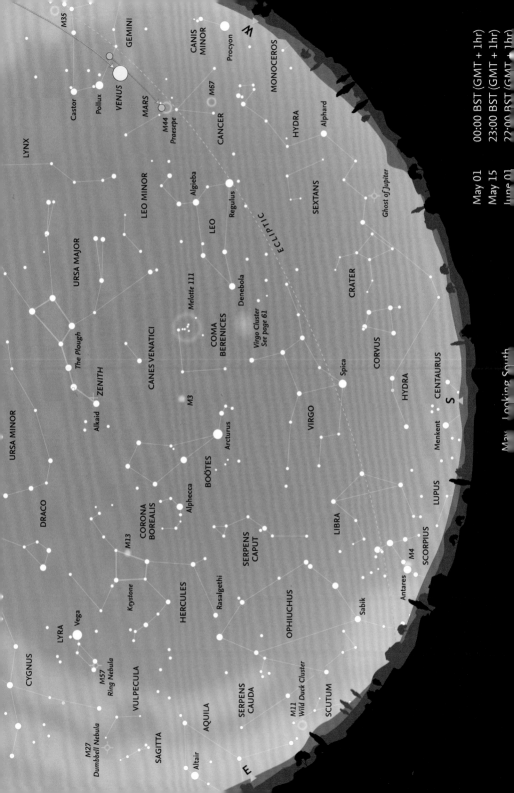

May Looking South

May 01   00:00 BST (GMT + 1hr)
May 15   23:00 BST (GMT + 1hr)
June 01   22:00 BST (GMT + 1hr)

May – Looking South

Early in the night, the constellation of **Virgo**, with **Spica** (α Virginis), lies due south, with **Leo** and both **Regulus** and **Denebola** (α and β Leonis, respectively) to its west still well clear of the horizon. Later in the night, the rather faint zodiacal constellation of **Libra** becomes visible and, to its east, the ruddy star **Antares** (α Scorpii) begins to climb up over the horizon.

Virgo contains the nearest large cluster of galaxies, which is the centre of the Local Supercluster, of which the Milky Way galaxy forms part. The Virgo Cluster contains some 2000 galaxies, the brightest of which are visible in amateur telescopes.

Arcturus in **Boötes** is high in the south, with the distinctive circlet of **Corona Borealis** clearly visible to its east. The brightest star (α Coronae Borealis) is known as **Alphecca**. The large constellation of **Ophiuchus** (which actually crosses the ecliptic, and is thus the 'thirteenth' zodiacal constellation) is climbing into the eastern sky. Before the constellation boundaries were formally adopted by the International Astronomical Union in 1930, the southern region of Ophiuchus was regarded as

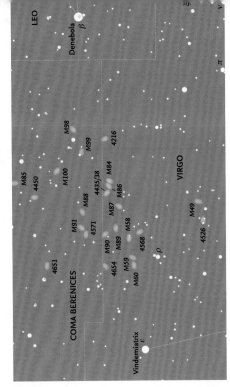

A finder chart for some of the brightest galaxies in the Virgo Cluster (see page 60). All stars brighter than magnitude 8.5 are shown.

forming part of the constellation of **Scorpius**, which had been part of the zodiac since antiquity.

The Moon's phases for May

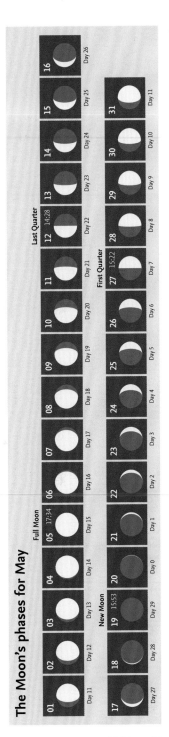

01 Day 11	02 Day 12	03 Day 13	04 Day 14	**Full Moon** 05 17:34 Day 15	06 Day 16	07 Day 17
08 Day 18	09 Day 19	10 Day 20	11 Day 21	**Last Quarter** 12 14:28 Day 22	13 Day 23	14 Day 24
15 Day 25	16 Day 26					

17 Day 27	18 Day 28	**New Moon** 19 15:53 Day 29	20 Day 0	21 Day 1	22 Day 2	23 Day 3
24 Day 4	25 Day 5	26 Day 6	**First Quarter** 27 15:22 Day 7	28 Day 8	29 Day 9	30 Day 10
31 Day 11						

May – Moon and Planets

The Moon

On May 4, one day before Full Moon, the Moon is 3.3° north of *Spica* (mag. 0.98) in *Virgo*. On May 5 there is a penumbral lunar eclipse. Invisible to the naked eye, mid-eclipse is over the Indian Ocean. Two days later, the Moon is 1.5° north of *Antares*. By May 13, one day after Last Quarter, the Moon is 3.3° south of *Saturn* (mag. 1.0) in *Aquarius* in the morning sky. It passes 0.8° north of *Jupiter* in *Pisces* on May 17, with the planet lost in the morning twilight. On May 19, at New Moon, it is 1.8° north of *Uranus* (mag. 5.9) in *Aries*. The Moon is 8.7° north of *Aldebaran* on May 20 and 2.2° north of *Venus* on May 23. It passes 1.6° south of *Pollux* on May 24, and, later the same day, 3.8° north of *Mars*. The Moon is 4.6° north of *Regulus* in *Leo* on May 27. It is again 3.3° north of *Spica* on May 31.

The planets

Mercury is lost in twilight throughout May. *Venus* is bright (mag. -4.2 to -4.4), and initially in *Taurus*, then crosses into *Gemini*. There is an occultation of the planet on May 27, but this is essentially invisible, occurring in daylight and only visible from a small area of the southern Indian Ocean. *Mars* (mag. 1.3 to 1.6) moves from *Gemini* into *Cancer*, in the evening sky. *Jupiter* is inside *Pisces*, lost in the morning twilight. *Saturn* (mag. 1.0) is in *Aquarius*. *Uranus* is in *Aries* at mag. 5.9 and comes to superior conjunction on May 9. *Neptune* remains in *Pisces* at mag. 7.9.

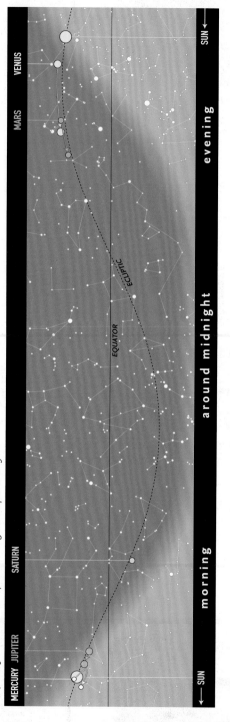

The path of the Sun and the planets along the ecliptic in May.

Calendar for May

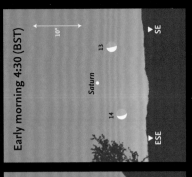

After midnight 2:00 (BST)

May 4 • The Moon is almost Full, when it passes Spica, in the southwest.

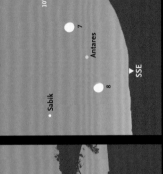

After midnight 1:00 (BST)

May 7–8 • The Moon passes Antares, in the south-southeast. Sabik is nearby.

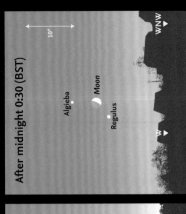

Early morning 4:30 (BST)

May 13–14 • The waning crescent Moon passes Saturn, low in the southeast.

Evening 22:00 (BST)

May 22–24 • In the evening twilight, the Moon passes Venus, Pollux & Castor and Mars. Alhena (γ Gem), Elnath (β Tau) and Menkalinan (β Aur) are nearby. Elnath is probably lost in dawn.

After midnight 0:30 (BST)

May 27 • The Moon is between Regulus and Algieba, low in the western sky.

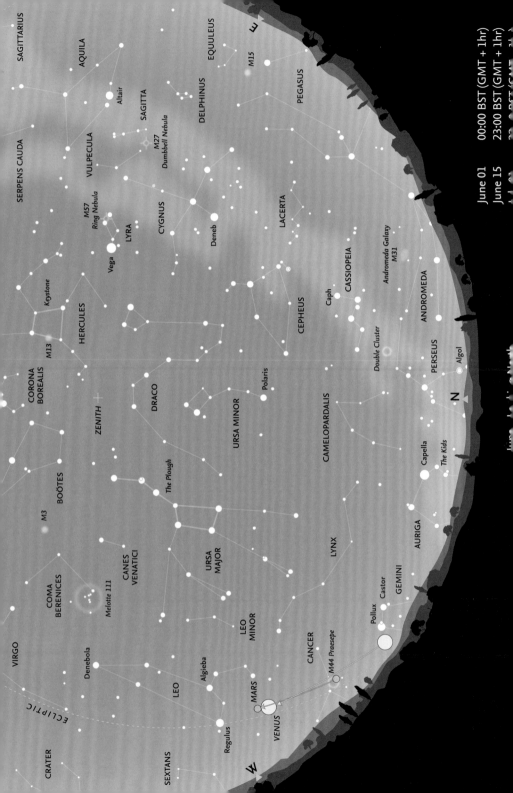

June – Looking North

Around summer solstice (June 21) even in southern England and Ireland a form of twilight persists throughout the night. Farther north, in Scotland, the sky remains so light that most of the fainter stars and constellations are invisible. There, even brighter stars, such as the seven stars making up the well-known asterism known as the **Plough** in **Ursa Major** may be difficult to detect except around local midnight, 00:00 UT (01:00 BST).

But there is one compensation during these light nights: even southern observers may be lucky enough to witness a display of noctilucent clouds (NLC). These are highly distinctive clouds shining with an electric-blue tint, observed in the sky in the direction of the North Pole. They are the highest clouds in the atmosphere, occurring at altitudes of 80–85 km, far above all other clouds. They are only visible during summer nights, for about a month or six weeks on either side of the solstice, when observers are in darkness, but the clouds themselves remain illuminated by sunlight, reaching them from the Sun, itself hidden below the northern horizon.

Noctilucent clouds, photographed by Alan Tough from Nairn in Scotland, on 31 May 2020, at 00:28.

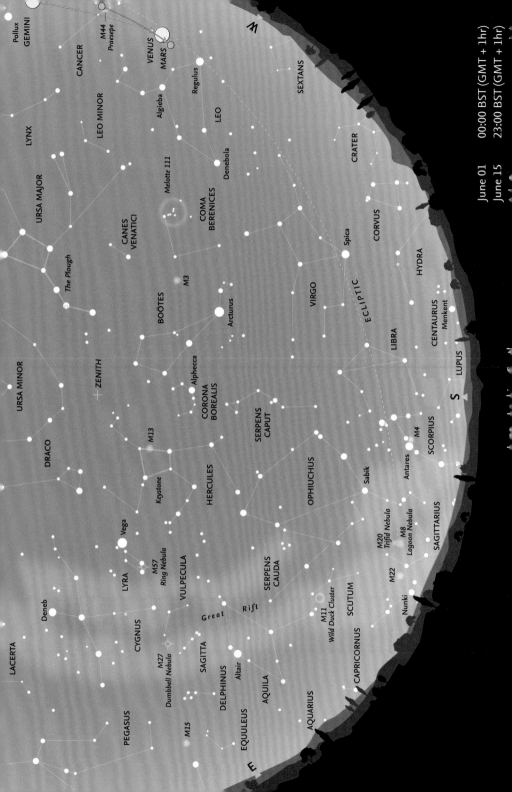

June 01   00:00 BST (GMT + 1hr)
June 15   23:00 BST (GMT + 1hr)

June – Looking South

Although the persistent twilight makes observing even the southern sky difficult, the rather undistinguished constellation of **Libra** lies almost due south. The red supergiant star **Antares** – the name means the 'Rival of Mars' – in **Scorpius** is visible slightly to the east of the meridian, but the 'tail' or 'sting' remains below the horizon. Higher in the sky is the large constellation of **Ophiuchus** (the 'Serpent Bearer'), lying between the two halves of the constellation of **Serpens: Serpens Caput** ('Head of the Serpent') to the west and **Serpens Cauda** ('Tail of the Serpent') to the east. (Serpens is the only constellation to be divided into two distinct parts.) The ecliptic runs across Ophiuchus, and the Sun actually spends far more time in the constellation than it does in the 'classical' zodiacal constellation of Scorpius, a small area of which lies between Libra and Ophiuchus.

Higher in the southern sky, the three constellations of **Boötes, Corona Borealis** and **Hercules** are now better placed for observation than at any other time of the year. This is an ideal time to observe the fine globular cluster of M13 in Hercules (see finder chart on page 59).

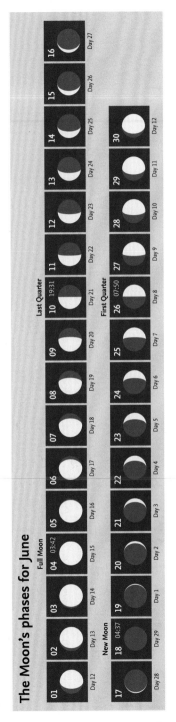

The constellations of Boötes and Corona Borealis are high in the sky in June. Arcturus has an orange tint and is the brightest star (mag. -0.05) in the northern celestial hemisphere.

The Moon's phases for June

			Full Moon						
01	02	03	04 03:42	05	06	07	08	09	Last Quarter 10 19:31
Day 12	Day 13	Day 14	Day 15	Day 16	Day 17	Day 18	Day 19	Day 20	Day 21
New Moon 17	18 04:37	19	20	21	22	23	24	25	First Quarter 26 07:50
Day 28	Day 29	Day 1	Day 2	Day 3	Day 4	Day 5	Day 6	Day 7	Day 8
11	12	13	14	15	16				
Day 22	Day 23	Day 24	Day 25	Day 26	Day 27				
27	28	29	30						
Day 9	Day 10	Day 11	Day 12						

June – Moon and Planets

The Moon

On June 3, one day before Full Moon, the Moon is 1.5° north of *Antares* in *Scorpius*. By June 9, waning gibbous and one day before Last Quarter, it is 2.0° south of *Saturn* (mag. 0.9) in *Aquarius*. By June 11, it is 2.0° south of *Neptune* (mag. 7.9) in *Pisces*. On June 14, the Moon is 1.5° north of *Jupiter*, in *Aries*, low in the morning sky. The next day, June 15, it passes 2.0° north of *Uranus* (mag. 5.8) also in Aries. On June 20, the waxing crescent Moon is 1.7° south of *Pollux* in *Gemini*. By June 22, the Moon is 3.7° north of brilliant *Venus* (mag. -4.6) in *Cancer*. Later the same day it passes 3.8° north of *Mars*. The next day, June 23, the Moon is 4.4° north of *Regulus* in *Leo*. By June 27, it is 3.1° north of *Spica* in *Virgo*.

The planets

Mercury is low in the morning twilight. *Venus*, moving steadily eastwards, brightens from mag. -4.4 to -4.7 over the month. *Mars*, at mag. 1.6 to 1.7, is in *Cancer*, but moves into *Leo* later in the month. *Jupiter* is in *Aries* at mag. -2.1 to -2.2. *Saturn* (mag. 0.9 to 0.8) is moving slowly in *Aquarius*. *Uranus* remains in *Aries* at mag. 5.8 and *Neptune* is in *Pisces* at mag. 7.9.

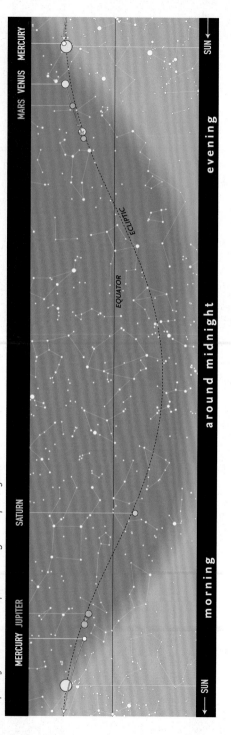

The path of the Sun and the planets along the ecliptic in June.

Calendar for June

03	21:53	Antares 1.5°S of the Moon
04	03:42	Full Moon
04	05:00 *	Uranus (mag. 5.8) 2.9°N of Mercury (mag. 0.1)
04	11:01	Venus at greatest elongation (45.4°E, mag. -4.4)
06	23:06	Moon at perigee = 364,861 km
09	20:23	Saturn 3.0°N of the Moon
10	19:31	Last Quarter
11	07:46	Neptune 2.0°N of the Moon
14	06:35	Jupiter 1.5°S of the Moon
15	09:54	Uranus 2.0°S of the Moon
16	20:38	Mercury 4.3°S of the Moon
16	23:09	Aldebaran 8.7°S of the Moon
18	04:37	New Moon
20	09:47	Pollux 1.7°N of the Moon
21	14:58	Summer solstice
22	00:48	Venus 3.7°S of the Moon
22	10:09	Mars 3.8°S of the Moon
22	18:30	Moon at apogee = 405,385 km
23	07:46	Regulus 4.4°S of the Moon
26	07:50	First Quarter
27	19:47	Spica 3.1°S of the Moon

* These objects are close together for an extended period around this time.

Evening 23:00 (BST)

June 3 • The almost Full Moon is close to Antares, with Sabik a little farther to the southeast.

Early morning 4:00 (BST)

June 10 • The Moon is almost at Last Quarter, when it is near Saturn, in the southeast.

Evening 22:15 (BST)

June 20–23 • In the evening, the narrow crescent Moon passes Venus, Mars, Regulus and Algieba. Pollux and Castor are probably lost in twilight, as is the very thin crescent of the Moon on June 20.

Early morning 3:45 (BST)

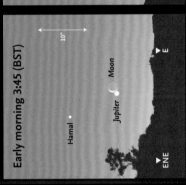

June 14 • Low in the east, Jupiter is close to the Moon. Hamal (α Ari) is about ten degrees higher.

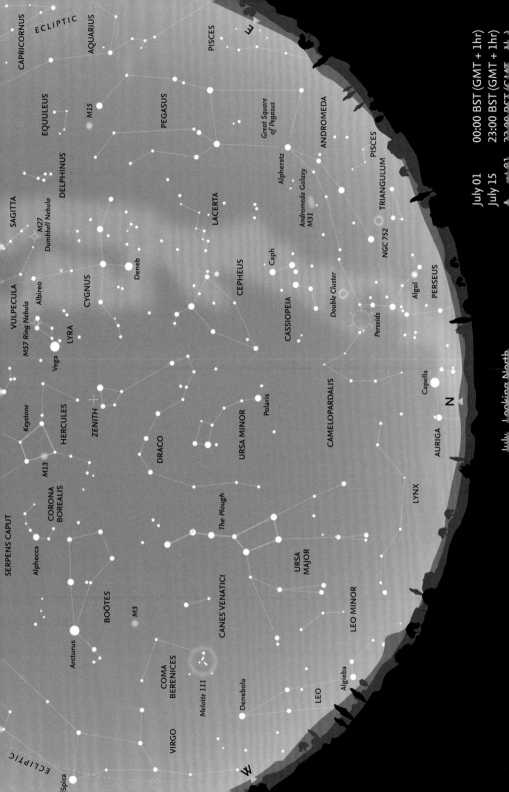

July, Looking North

July 01   00:00 BST (GMT + 1hr)
July 15   23:00 BST (GMT + 1hr)

July – Looking North

As in June, light nights and the chance of observing noctilucent clouds persist throughout July, but later in the month (and particularly after midnight) some of the major constellations begin to be more easily seen. **Capella**, the brightest star in **Auriga** (most of which is too low to be visible), is skimming the northern horizon. **Cassiopeia** is clearly visible in the northeast and **Perseus**, to its south, is beginning to climb clear of the horizon. The band of the Milky Way, from Perseus through Cassiopeia towards **Cygnus**, stretches up into the northeastern sky. If the sky is dark and clear, you may be able to make out the small, faint constellation of **Lacerta**, lying across the Milky Way between Cassiopeia and Cygnus. In the east, the stars of **Pegasus** are now well clear of the horizon, with the main line of stars forming **Andromeda** roughly parallel to the horizon in the northeast. **Alpheratz** (α Andromedae) is actually the star at the northeastern corner of the **Great Square of Pegasus. Cepheus** and **Ursa Major** are on opposite sides of **Polaris** and **Ursa Minor,** in the east and west, respectively. The head of **Draco** is very close to the zenith so the whole of this winding constellation is readily seen.

Meteors

July brings increasing meteor activity, mainly because there are several minor radiants active in the constellations of **Capricornus** and **Aquarius.** Because of their location, however, observing conditions are not particularly favourable for northern-hemisphere observers, although the first shower, the **α-Capricornids,** active from July 3 to August 15 (peaking July 30, with a tail to August 15), does often produce very bright fireballs. The maximum rate, however, is only about 5 per hour. The parent body is Comet 169P/NEAT. The most prominent shower is probably that of the **Southern δ-Aquariids,** which are active from around July 12 to August 23, with a peak on July 30, although even then the rate is unlikely to reach 25 meteors per hour. In this

Cygnus, sometimes known as the 'Northern Cross', depicts a swan 'flying' down the Milky Way towards Sagittarius. The brightest star, Deneb (α Cygni), represents the tail, and Albireo (β Cygni) marks the position of the head, and lies near the bottom of the image.

case, the parent body may be Comet 96P/Machholz. This year, both shower maxima occur when the Moon is two days before full, so observing conditions are particularly unfavourable. A chart showing the **δ-Aquariids** radiant is shown on page 30. The **Perseids** begin on July 17 and peak on August 12–13.

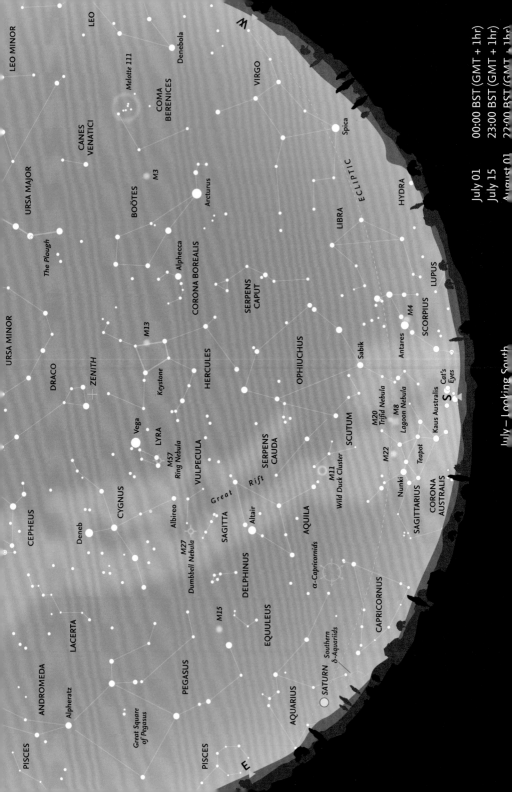

July – Looking South

July 01	00:00 BST (GMT + 1hr)
July 15	23:00 BST (GMT + 1hr)
August 01	22:00 BST (GMT + 1hr)

LEO MINOR
LEO
URSA MAJOR
CANES VENATICI
Melotte 111
COMA BERENICES
Denebola
M3
BOÖTES
The Plough
Arcturus
VIRGO
Spica
Alphecca
CORONA BOREALIS
SERPENS CAPUT
LIBRA
ECLIPTIC
HYDRA
URSA MINOR
DRACO
M13
HERCULES
Keystone
ZENITH
OPHIUCHUS
Sabik
LUPUS
M4
SCORPIUS
Antares
Cat's Eyes
Kaus Australis
S
Vega
LYRA
M57 Ring Nebula
CYGNUS
Deneb
CEPHEUS
VULPECULA
SERPENS CAUDA
SCUTUM
M11 Wild Duck Cluster
Great Rift
M20 Trifid Nebula
M8 Lagoon Nebula
Teapot
M22
Nunki
SAGITTARIUS
CORONA AUSTRALIS
Albireo
SAGITTA
Altair
AQUILA
M27 Dumbbell Nebula
DELPHINUS
α-Capricornids
LACERTA
ANDROMEDA
Alpheratz
PEGASUS
M15
EQUULEUS
CAPRICORNUS
Great Square of Pegasus
Southern δ-Aquariids
SATURN
AQUARIUS
PISCES
E

July – Looking South

Although part of the constellation remains hidden, this is perhaps the best time of year to see **Scorpius**, with deep red **Antares** (α Scorpii), glowing just above the southern horizon. At around midnight (UT), 01:00 BST, part of **Sagittarius**, with the distinctive asterism of the 'Teapot', and the dense star clouds of the centre of the Milky Way, are just visible in the south. The **Great Rift** – actually dust clouds that hide the more distant stars – runs down the Milky Way from **Cygnus** towards Sagittarius. Towards its northern end is the small constellation of **Sagitta** and the planetary nebula **M27** (the Dumbbell Nebula) in the otherwise insignificant constellation of **Vulpecula**. The sprawling constellation of **Ophiuchus** lies close to the meridian for a large part of the month, separating the two halves of the constellation of **Serpens**. The western half is called **Serpens Caput** (Head of the Serpent) and the eastern part **Serpens Cauda** (Tail of the Serpent). In the east, the bright 'Summer Triangle', consisting of **Vega** in **Lyra**, **Deneb** in Cygnus and **Altair** in **Aquila**, begins to dominate the southern sky, as it will throughout August and into September. The small constellation of Lyra, with Vega and a distinctive quadrilateral of stars to its east and south, lies not far south of the zenith.

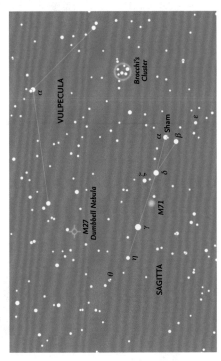

A finder chart for M27, the Dumbbell Nebula, a relatively bright (magnitude 8) planetary nebula – a shell of material ejected in the late stages of a star's lifetime – in the constellation of Vulpecula. All stars brighter than magnitude 7.5 are shown.

The Moon's phases for July

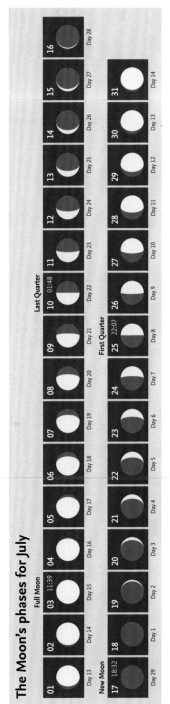

July – Moon and Planets

The Earth

The Earth reaches aphelion, the farthest point from the Sun in its yearly orbit, at 20:07 on July 6 at a distance of 1.016680584 AU = 152,093,178.67 km.

The Moon

The Moon is 1.5° north of **Antares** in **Scorpius** on July 1, two days before Full Moon. On July 7, four days after Full, it is 2.7° south of **Saturn** (mag. 0.8) in **Aquarius.** By July 11, the Moon is 2.2° north of **Jupiter** (mag. -2.3) in **Aries.** On July 12 the waning crescent is 2.3° north of **Uranus** (mag. 5.8). By July 14 it is 8.8° north of **Aldebaran.** Three days later, at New Moon, it is 1.8° south of **Pollux** in **Gemini.** On July 19 it passes 3.5° north of **Mercury** (mag. -0.5), low in the evening twilight.

The next day the waxing crescent is 7.9° north of brilliant **Venus** (mag. -4.6), also low in the evening sky. Later that day, the Moon is 4.2° north of **Regulus,** between it and **Algieba** (γ Leonis). On July 21 the waxing crescent is 3.3° north of **Mars,** also in **Leo.** Before First Quarter on July 25 it is 2.8° north of **Spica** in **Virgo.** By July 28 it is 1.2° north of **Antares.**

The planets

Mercury, reaches superior conjunction on July 1, but is too close to the Sun to be readily visible. **Venus** is very bright (mag. -4.7 to -4.4) so may be visible although close to the Sun. **Mars** is mag. 1.7 to 1.8 in **Leo.** **Jupiter** is mag. -2.3 in **Aries. Saturn** is in **Aquarius. Uranus** at mag. 5.8 is in **Aries** and is on the border of **Taurus** at the end of the month. **Neptune** remains in **Pisces** at mag. 7.9.

The path of the Sun and the planets along the ecliptic in July.

Calendar for July

01	05:06	Mercury at superior conjunction
01	07:55	Antares 1.5°S of the Moon
03–Aug 15		α-Capricornid meteor shower
03	11:39	Full Moon
04	22:25	Moon at perigee = 360,149 km
06	20:07	Earth at aphelion (1.016680584 AU)
07	03:10	Saturn 2.7°N of the Moon
07	19:58	Minor planet (15) Eunomia at opposition (mag. 8.7)
08	14:12	Neptune 1.7°N of the Moon
10	01:48	Last Quarter
11	21:21	Jupiter 2.2°S of the Moon
12–Aug 23		Southern δ-Aquariid meteor shower
12	17:48	Uranus 2.3°S of the Moon
14	05:05	Aldebaran 8.8°S of the Moon
17–Aug 24		Perseid meteor shower
17	16:22	Pollux 1.8°N of the Moon
17	18:32	New Moon
19	08:56	Mercury 3.5°S of the Moon
20	06:57	Moon at apogee = 406,289 km
20	08:37	Venus 7.9°S of the Moon
20	14:32	Regulus 4.2°S of the Moon
21	04:00	Mars 3.3°S of the Moon
25	03:41	Spica 2.8°S of the Moon
25	22:07	First Quarter
26	13:00 *	Mercury (mag. −0.2) 5.3°N of Venus (mag. −4.5)
28	17:46	Antares 1.2°S of the Moon
30		α-Capricornid meteor shower maximum
30		Southern δ-Aquariid meteor shower maximum

* These objects are close together for an extended period around this time.

After midnight 0:30 (BST)

July 1–2 • *The Moon passes Antares around midnight. Sabik is about 10 degrees higher.*

Early morning 4:00 (BST)

July 7 • *The Moon is close to Saturn. Fomalhaut is closer to the horizon.*

Early morning 4:00 (BST)

July 11–13 • *The Moon passes Jupiter, Uranus (mag. 5.8) and the Pleiades.*

Evening 21:45 (BST)

July 20–21 • *The Moon passes Regulus, Algieba and Mars (mag. 1.7). Denebola (β Leo) is higher in the west.*

Evening 21:30 (BST)

July 28 • *Almost due south, the Moon and Antares are joined up again, with Sabik nearby.*

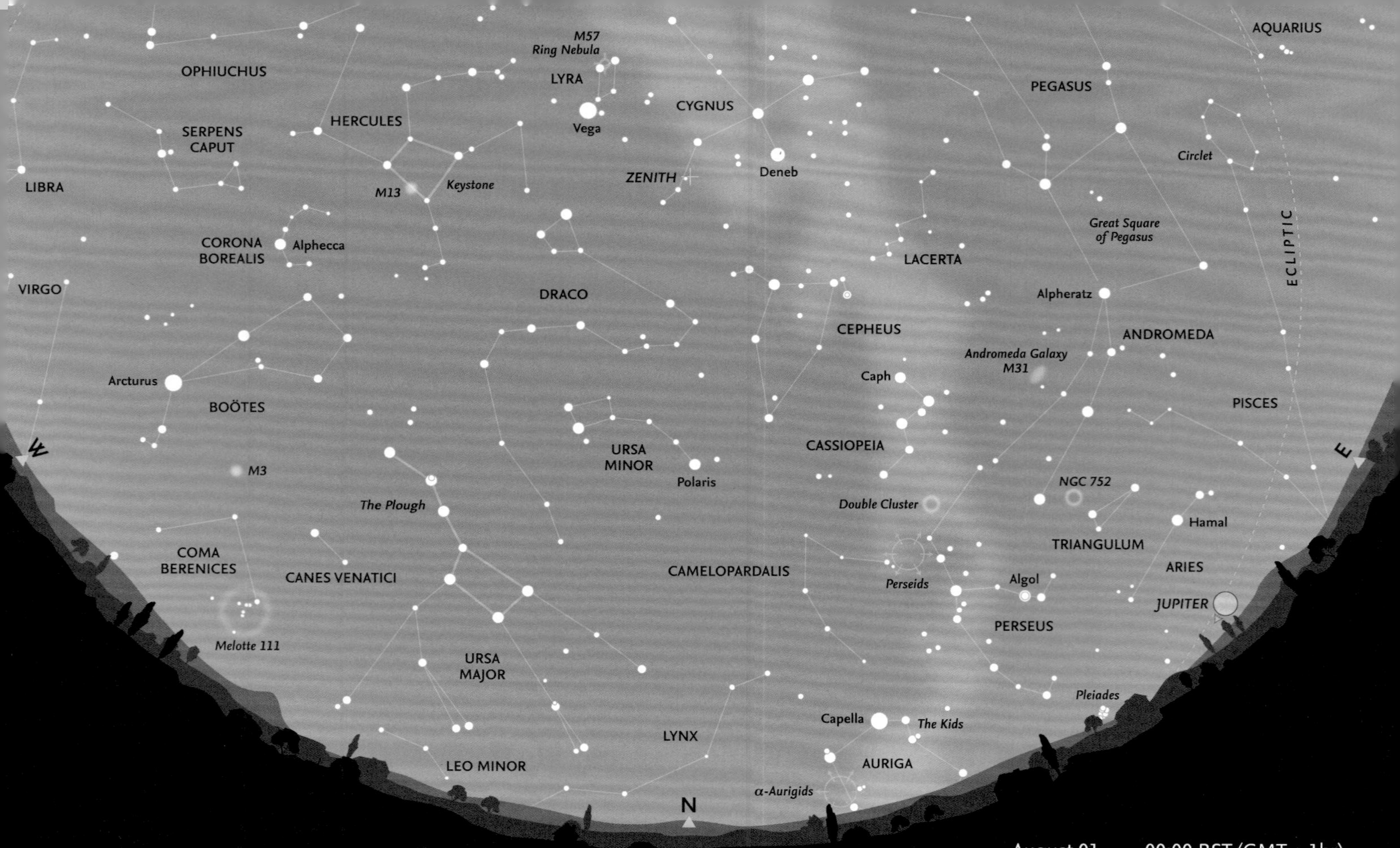

August – Looking North

August 01	00:00 BST (GMT + 1hr)
August 15	23:00 BST (GMT + 1hr)
September 01	22:00 BST (GMT + 1hr)

Constellations and features labelled on the chart:

AQUARIUS
ECLIPTIC
PISCES
Circlet
ANDROMEDA
Alpheratz
Great Square of Pegasus
PEGASUS
Hamal
ARIES
JUPITER
NGC 752
TRIANGULUM
Andromeda Galaxy M31
Algol
PERSEUS
LACERTA
Caph
Perseids
Pleiades
CASSIOPEIA
Double Cluster
The Kids
CEPHEUS
Deneb
AURIGA
Capella
CYGNUS
CAMELOPARDALIS
α-Aurigids
ZENITH
M57 Ring Nebula
LYRA
Vega
Polaris
DRACO
URSA MINOR
LYNX
N
Keystone
HERCULES
M13
URSA MAJOR
LEO MINOR
Alphecca
CORONA BOREALIS
The Plough
OPHIUCHUS
SERPENS CAPUT
BOÖTES
M3
CANES VENATICI
Melotte 111
COMA BERENICES
Arcturus
LIBRA
VIRGO
Arcturus
W
E

August – Looking North

Ursa Major is now the 'right way up' in the northwest, although some of the fainter stars in the south of the constellation are difficult to see. Beyond it, *Boötes* stands almost vertically in the west, but pale orange *Arcturus* is sinking towards the horizon. Higher in the sky, both *Corona Borealis* and *Hercules* are clearly visible.

In the northeast, *Capella* is clearly seen, but most of *Auriga* still remains below the horizon. Higher in the sky, *Perseus* is gradually coming into full view and, later in the night and later in the month, the beautiful *Pleiades* cluster rises above the northeastern horizon. Between Perseus and *Polaris* lies the faint and unremarkable constellation of *Camelopardalis*.

Higher still, both *Cassiopeia* and *Cepheus* are well placed for observation, despite the fact that Cassiopeia is completely immersed in the band of the Milky Way, as is the 'base' of Cepheus. *Pegasus* and *Andromeda* are now well above the eastern horizon and, below them, the constellation of *Pisces* is climbing into view. Two of the stars in the 'Summer Triangle', *Deneb* and *Vega*, are close to the zenith high overhead

Meteors

August is the month when one of the best meteor showers of the year occurs: the *Perseids*. This is a long shower, generally beginning about July 17 and continuing until around August 24, with a maximum on August 12–13, when the rate may reach as high as 100 meteors per hour (and on rare occasions, even higher). In 2023, maximum is about three days before New Moon, so conditions are favourable. The Perseids are debris from Comet 109P/Swift-Tuttle (the Great Comet of 1862). Perseid meteors are fast and many of the brighter ones leave persistent trains. Some bright fireballs also occur during the shower.

A brilliant Perseid fireball, streaking alongside the Great Rift in the Milky Way, photographed in 2012 by Jens Hackmann from near Weikersheim in Germany. Four additional, fainter Perseids are also visible in the image.

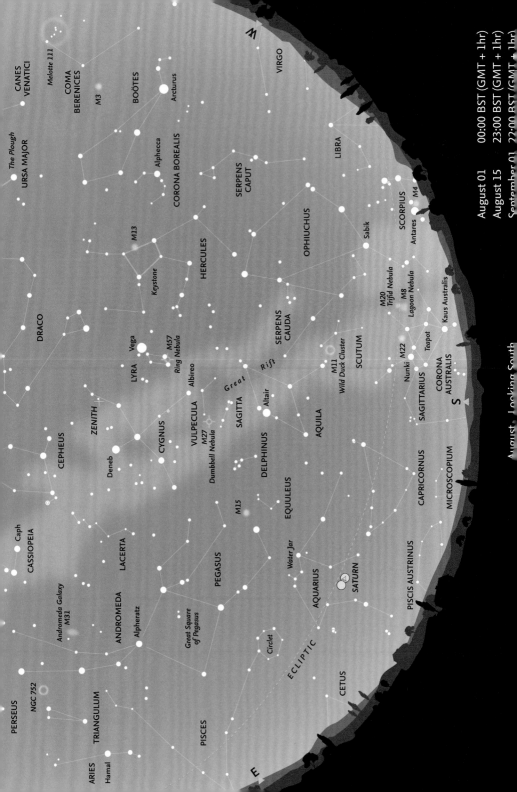

August 01   00:00 BST (GMT + 1hr)
August 15   23:00 BST (GMT + 1hr)
September 01 22:00 BST (GMT + 1hr)

August – Looking South

August – Looking South

The whole stretch of the summer Milky Way stretches across the sky in the south, from **Cygnus**, high in the sky near the zenith, past **Aquila**, with bright **Altair** (α Aquilae), to part of the constellation of **Sagittarius** close to the horizon, where the pattern of stars known as the 'Teapot' may be just visible. This area contains many nebulae and both open and globular clusters. Between **Albireo** (β Cygni) and Altair lie the two small constellations of **Vulpecula** and **Sagitta**, with the latter easier to distinguish (because of its shape) from the clouds of the Milky Way. Between Sagitta and **Pegasus** to the east lie the highly distinctive five stars that form the tiny constellation of **Delphinus** (again, one of the few constellations that actually bear some resemblance to the creatures after which they are named). Below Aquila, mainly in the star clouds of the Milky Way, lies **Scutum**, most famous for the bright open cluster, **M11** or the 'Wild Duck Cluster', readily visible in binoculars. To the southeast of Aquila lie the two zodiacal constellations of **Capricornus** and **Aquarius**.

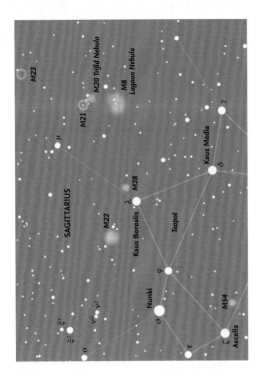

A finder chart for the gaseous nebulae M8 (the Lagoon Nebula), M20 (the Trifid Nebula) and the globular cluster M22, all in Sagittarius. Clusters M21, M23 (open) and M28 (globular) are faint. The chart shows all stars brighter than magnitude 7.5.

The Moon's phases for August

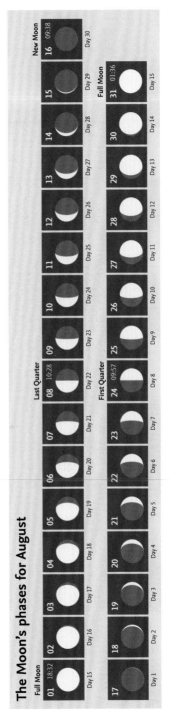

August – Moon and Planets

The Moon

On August 3, just after Full Moon, it is 2.5° south of *Saturn* in *Aquarius*. By August 8, waning gibbous, it is 2.9° north of *Jupiter* in *Pisces*. On August 9 it is 2.6° north of *Uranus* (mag. 5.8), close to *Taurus*. The next day, it is 9.0° north of *Aldebaran*. On August 13, a waning crescent, it is 1.7° south of *Pollux* in *Gemini*. On August 15, it is 13.3° north of *Venus* (mag. -4.1) in *Pisces* in the early evening sky. At New Moon on August 16, it is 4.1° north of *Regulus*. On August 18 it is 2.2° north of *Mars* in *Virgo*. On August 21 it is 2.5° north of *Spica* on the edge of evening twilight. On August 25, one day after First Quarter, it is 1.1° north of *Antares* in *Scorpius*. By August 30, one day before Full Moon, it is again 2.5° south of *Saturn* in *Aquarius*.

The Planets

Mercury is low in the twilight. It comes to greatest elongation (27.4°) east at mag. -0.3, on August 10. *Venus* (initially mag. -4.3, fades to mag. -4.1, then brightens to mag. -4.6) may be detectable in morning twilight. *Mars* (mag. 1.8) may be visible at the beginning of evening twilight. *Jupiter* is clearly visible in *Aries* at mag. -2.4 to -2.6. *Saturn*, in *Aquarius*, is mag. 0.4 at opposition on August 27. *Uranus* remains in *Aries* at mag. 5.9 and *Neptune* in *Pisces* at mag. 7.8.

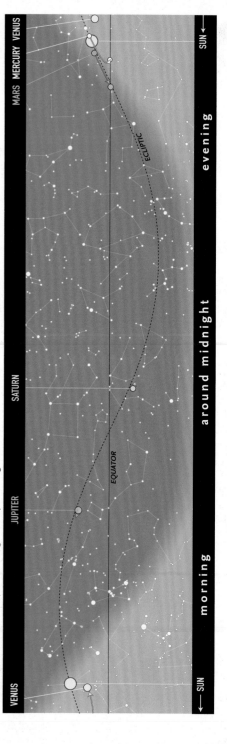

The path of the Sun and the planets along the ecliptic in August.

Calendar for August

01	18:32	Full Moon
02	05:52	Moon at perigee = 357,311 km
03	10:26	Saturn 2.5°N of the Moon
04	22:02	Neptune 1.5°N of the Moon
08	09:44	Jupiter 2.9°S of the Moon
08	10:28	Last Quarter
09	01:03	Uranus 2.6°S of the Moon
10	01:47	Mercury at greatest elongation (27.4°E, mag. –0.3)
10	10:43	Aldebaran 9.0°S of the Moon
12–13		Perseid meteor shower maximum
13	11:15	Venus at inferior conjunction
13	22:14	Pollux 1.79°N of the Moon
15	17:00	Venus 13.3°S of the Moon
16	09:38	New Moon
16	11:54	Moon at apogee = 406,635 km (most distant of year)
16	20:37	Regulus 4.1°S of the Moon
18	11:26	Mercury 6.9°S of the Moon
18	23:07	Mars 2.2°S of the Moon
21	10:08	Spica 2.5°S of the Moon
24	09:57	First Quarter
25	02:06	Antares 1.1°S of the Moon
27	07:39	Minor planet (8) Flora at opposition (mag. 8.3)
27	08:28	Saturn at opposition (mag. 0.4)
28–Sep.05		α-Aurigid meteor shower
30	15:54	Moon at perigee = 357,181 km (closest of year)
30	18:08	Saturn 2.5°N of the Moon
31	01:36	Full Moon

Early morning 4:30 (BST)

August 3–4 • The Moon passes Saturn, one hour before sunrise. Fomalhaut is closer to the horizon.

Early morning 4:30 (BST)

August 8–9 • The Moon passes Jupiter, Uranus (mag. 5.8) and the Pleiades. Aldebaran is nearby.

Morning 5:00 (BST)

August 13–14 • The narrow crescent Moon passes Pollux and Castor, shortly before sunrise.

Evening 21:00 (BST)

August 24–25 • The Moon passes Antares, low in the south-southwest. Sabik is about twelve degrees higher.

Evening 21:30 (BST)

August 30 • The Moon is almost Full, when it joins Saturn again, low in the southeast.

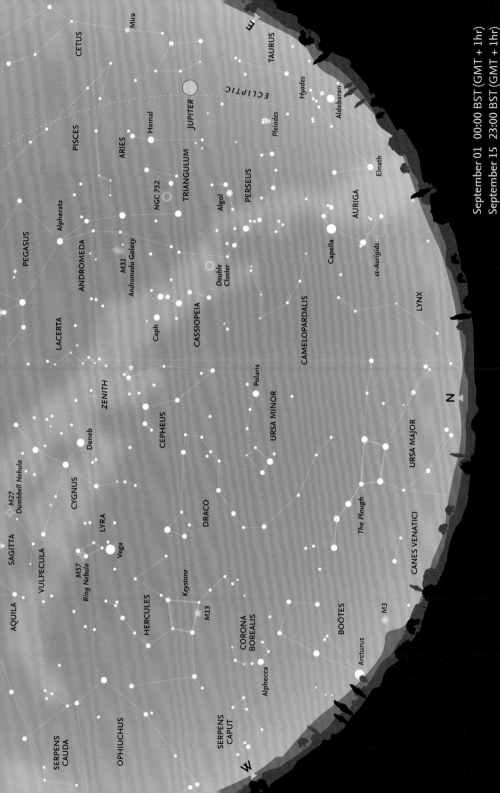

September – Looking North

The (northern) autumnal equinox occurs on September 23, when the Sun moves south of the equator in Virgo.

Ursa Major is now low in the north and to the northwest **Arcturus** and much of **Boötes** sink below the horizon later in the night and later in the month. In the northeast, **Auriga** is beginning to climb higher in the sky. Later in the month, **Taurus**, with orange **Aldebaran** (α Tauri), and even **Gemini**, with **Castor** and **Pollux**, become visible in the east and northeast. Due east, **Andromeda** is now clearly visible, with the small constellations of **Triangulum** and **Aries** (the latter a zodiacal constellation) directly below it. Practically the whole of the northern Milky Way is visible, arching across the sky, both in the north and in the south. It is not particularly clear in Auriga, or even **Perseus**, but in **Cassiopeia** and on towards **Cygnus** the clouds of stars become easier to see. The **Double Cluster** in Perseus is well placed for observation. **Cepheus** is 'upside-down' near the zenith, and the head of **Draco** and **Hercules** (with the globular cluster **M13**), beyond it are well placed for observation.

Meteors

After the major Perseid shower in August, there is very little shower activity in September. One minor, but very extended, shower, known as the **α-Aurigids**, tends to have two peaks of activity. The principal peak occurs on September 1. In 2023, the Moon is one day after Full, so conditions are very unfavourable. At maximum, however, the hourly rate hardly reaches 10 meteors per hour, although the meteors are bright and relatively easy to photograph. Activity from this shower may even extend into October. The **Southern Taurid** shower begins this month (on September 10) and, although rates are low, often produces very bright fireballs. This is a very long shower, lasting until about November 20. As a slight compensation for the lack of shower activity, however, in September the number of sporadic meteors reaches its highest rate at any time during the year.

The twin open clusters, known as the Double Cluster, in Perseus (more formally called h and χ Persei) are close to the main portion of the Milky Way.

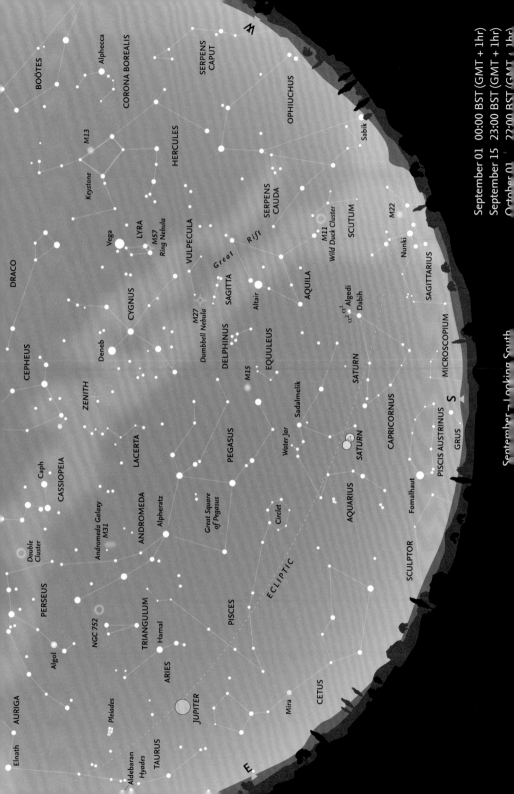

September 01 00:00 BST (GMT + 1hr)
September 15 23:00 BST (GMT + 1hr)
October 01 22:00 BST (GMT + 1hr)

September – Looking South

September – Looking South

The 'Summer Triangle' is now high in the southwest, with the Great Square of **Pegasus** high in the southeast. Below Pegasus are the two zodiacal constellations of **Capricornus** and **Aquarius**. In what is otherwise an unremarkable constellation, **Algedi** (α Capricorni) is actually a visual binary, with the two stars (α¹ Cap and α² Cap) readily seen with the naked eye. **Dabih** (β Capricorni), just to the south, is also a double star, and the components are relatively easy to separate with binoculars. In Aquarius, just to the east of **Sadalmelik** (α Aquarii) there is a small asterism consisting of four stars, resembling a tiny letter 'Y', known as the 'Water Jar'. Below Aquarius is a sparsely populated area of the sky with just one bright star in the constellation of **Piscis Austrinus**. In classical illustrations, water is shown flowing from the 'Water Jar' towards bright **Fomalhaut** (α Piscis Austrini).

Another zodiacal constellation, **Pisces**, is now clearly visible to the east of Aquarius. Although faint, there is a distinctive asterism of stars, known as the 'Circlet', south of the Great Square and another line of faint stars to the east of Pegasus. Still farther down towards the horizon is the constellation of **Cetus**, with the famous variable star **Mira** (ο Ceti) at its

The constellations of Pisces and Aries. The 'Circlet' of Pisces is bottom right and the three main stars of Aries, top left. In 2023, Jupiter is in Aries from June until December.

centre. When Mira is at maximum brightness (around mag. 3.5) it is clearly visible to the naked eye, but it disappears as it fades towards minimum (about mag. 9.5 or less). There is a finder chart for Mira on page 97.

The Moon's phases for September

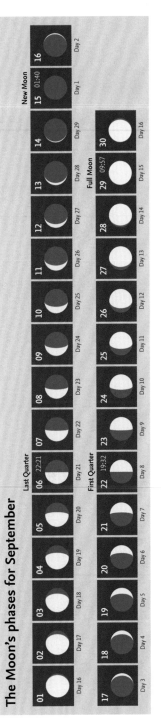

September – Moon and Planets

The Moon

On September 4, the waning gibbous Moon is 3.3° north of *Jupiter* (mag. -2.6) in *Aries*. The next day, it is 2.8° north of *Uranus* in *Aries*, which is much fainter at mag. 5.8. By September 6, at Last Quarter, it is 9.3° north of *Aldebaran* in *Taurus*. On September 10, the Moon is 1.6° south of *Pollux*. On September 11, it is 11.4° north of brilliant *Venus* at mag. –4.7 in *Cancer* in the morning sky. On September 13, it is 4.1° north of *Regulus* (mag. 1.4), which may just be visible in morning twilight. By September 16, one day after New Moon, it is 0.7° north of *Mars* in *Virgo*, at the end of evening twilight. The next day it is 2.4° north of *Spica*. By September 21, the Moon is 0.9° north of *Antares* in *Scorpius*. On September 27, the waxing gibbous Moon is 2.7° south of *Saturn* (mag. 0.5) in *Aquarius*.

The Planets

Mercury is generally lost in morning twilight, but comes to greatest western elongation (17.9°) at mag. -0.5 on September 22. *Venus* is in *Cancer*, visible in the morning sky at mag. –4.7. *Mars*, in *Virgo*, is lost in evening twilight. *Jupiter* begins retrograde motion in early September. It is in Aries at mag. –2.6 to -2.8. *Saturn* (mag. 0.5) is in *Aquarius*, and still retrograding. Uranus is just inside *Aries* at mag. 5.9 and *Neptune* is in *Pisces* at mag. 7.8 and comes to opposition on September 19.

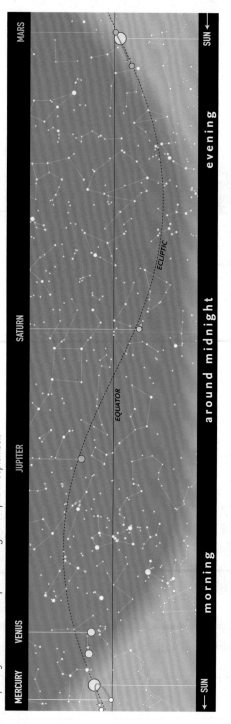

The path of the Sun and the planets along the ecliptic in September.

Calendar for September

	Time	Event
1		α-Aurigid meteor shower maximum
1	07:21	Neptune 1.4°N of the Moon
4	19:47	Jupiter 3.3°S of the Moon
5	08:45	Uranus 2.8°S of the Moon
6	11:09	Mercury at inferior conjunction
6	17:21	Aldebaran 9.3°S of the Moon
6	22:21	Last Quarter
10–Nov.20		Southern Taurid meteor shower
0	04:10	Pollux 1.6°N of the Moon
1	12:59	Venus 11.4°S of the Moon
2	15:43	Moon at apogee = 406,291 km
3	02:39	Regulus 4.1°S of the Moon
3	17:40	Mercury 6.0°S of the Moon
5	01:40	New Moon
6	19:20	Mars 0.7°S of the Moon
7	15:51	Spica 2.4°S of the Moon
9	11:17	Neptune at opposition (mag. 7.8)
1	08:27	Antares 0.9°S of the Moon
2	13:16	Mercury at greatest elongation (17.9°W, mag. -0.5)
2	19:32	First Quarter
3	06:50	Autumnal equinox
7	01:29	Saturn 2.7°N of the Moon
8	00:59	Moon at perigee = 359,911 km
8	16:59	Neptune 1.4°N of the Moon
9	09:57	Full Moon

After midnight 0:30 (BST)

September 5–6 • The Moon passes Jupiter, Uranus (mag. 5.7) and the Pleiades. Aldebaran is closer to the horizon.

Morning 5:00 (BST)

September 10–12 • The waning crescent Moon passes Pollux and Castor and is near Venus two days later. Alhena and Procyon are farther to the southeast.

Evening 20:00 (BST)

September 20–21 • The Moon is moving close to the horizon, when it passes Antares. On September 21, Sabik is about twelve degrees above it.

After midnight 1:00 (BST)

September 27 • Saturn is straight above the waxing gibbous Moon. Fomalhaut (α PsA), the brightest star in this area is closer to the horizon.

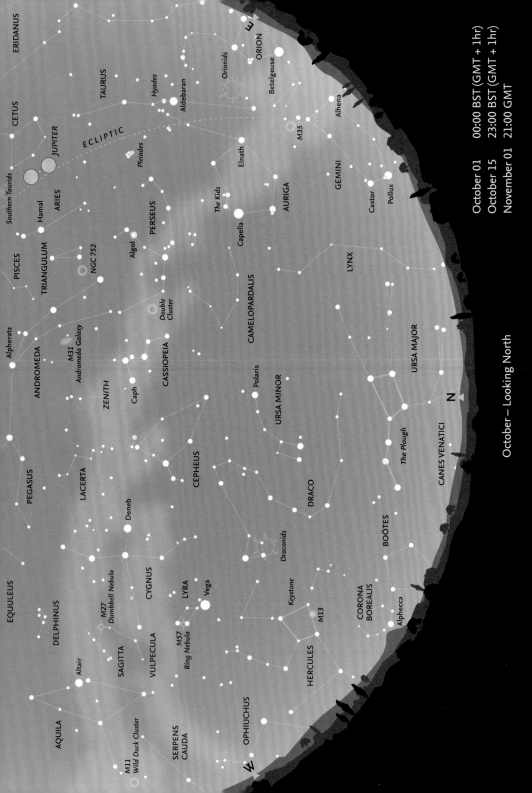

October – Looking North

October 01   00:00 BST (GMT + 1hr)
October 15   23:00 BST (GMT + 1hr)
November 01   21:00 GMT

October – Looking North

Ursa Major is grazing the horizon in the north, while high overhead are the constellations of **Cepheus**, **Cassiopeia** and **Perseus**, with the Milky Way between Cepheus and Cassiopeia near the zenith. **Auriga** is now clearly visible in the east, as is **Taurus** with the **Pleiades**, **Hyades** and orange **Aldebaran**. Also in the east, **Orion** and **Gemini** are starting to rise clear of the horizon.

The constellations of **Boötes** and **Corona Borealis** are now essentially lost to view in the northwest, and **Hercules** is also descending towards the western horizon. The three stars of the Summer Triangle are still clearly visible, although **Aquila** and **Altair** are beginning to approach the horizon in the west. Towards the end of the month (October 29) Summer Time ends in Europe, with Britain reverting to Greenwich Mean Time and Europe to Central European Time.

Meteors

The **Orionids** are the major, fairly reliable meteor shower active in October. Like the May **η-Aquariid** shower, the Orionids are associated with Comet 1P/Halley. During this second pass through the stream of particles from the comet, slightly fewer meteors are seen than in May, but conditions are more favourable for northern observers. In both showers the meteors are very fast, and many leave persistent trains. Although the Orionid maximum is quoted as October 21, in fact there is a very broad maximum, lasting about a week roughly centred on that date, with hourly rates around 25. Occasionally, rates are higher (50–70 per hour). In 2023, the Moon is around First Quarter, so conditions are reasonably favourable.

The faint shower of the **Southern Taurids** (often with bright fireballs) peaks on October 10. The Southern Taurid maximum occurs when the Moon is a waning crescent, so conditions are more favourable than

The constellation of Perseus is not only the location of the radiant of the Perseid meteor shower in August, but is also well known for the pair of open clusters, called the Double Cluster (near the top edge of the image), close to the border with Cassiopeia, and also for Algol (β Persei), the famous variable star (just below the centre of the photograph).

those for the Orionids. Towards the end of the month (around October 20), another shower (the Northern Taurids) begins to show activity, which peaks early in November. The parent comet for both Taurid showers is Comet 2P/Encke. The meteors in both Taurid streams are relatively slow and bright.

A minor shower, the **Draconids**, begins on October 6 and peaks on October 8–9.

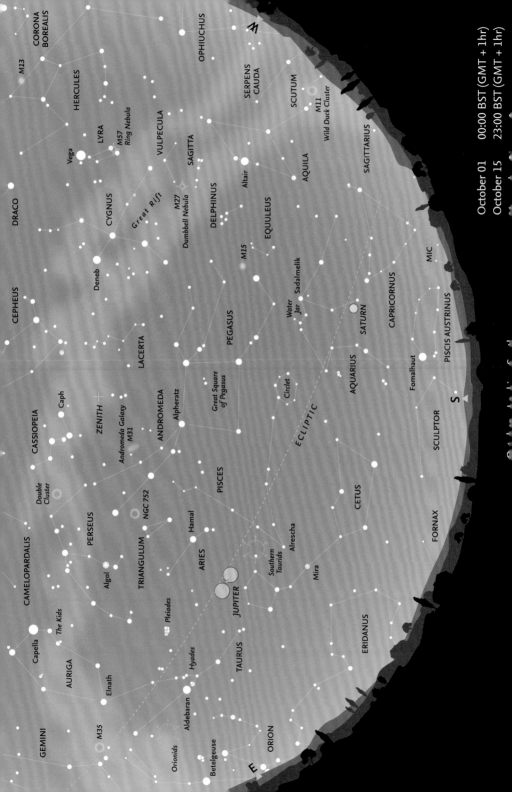

October 01   00:00 BST (GMT + 1hr)
October 15   23:00 BST (GMT + 1hr)

October – Looking South

The Great Square of **Pegasus** dominates the southern sky, framed by the two chains of stars that form the constellation of **Pisces**, together with **Alrescha** (α Piscium) at the point where the two lines of stars join. Also clearly visible is the constellation of **Cetus**, below Pegasus and Pisces. Although **Capricornus** is beginning to disappear, **Aquarius** to its east is well placed in the south, with solitary **Fomalhaut** and the constellation of **Piscis Austrinus** beneath it, close to the horizon.

The main band of the Milky Way and the Great Rift runs down from **Cygnus**, through **Vulpecula**, **Sagitta** and **Aquila** towards the western horizon. **Delphinus** and the tiny, unremarkable constellation of **Equuleus** lie between the band of the Milky Way and Pegasus. **Andromeda** is clearly visible high in the sky to the southeast, with the small constellation of **Triangulum** and the zodiacal constellation of **Aries** below it. **Perseus** is high in the east, and by now the **Pleiades** and **Taurus** are well clear of the horizon. Later in the night, and later in the month, **Orion** rises in the east, a sign that the autumn season has arrived and of the steady approach of winter.

The Moon's phases for October

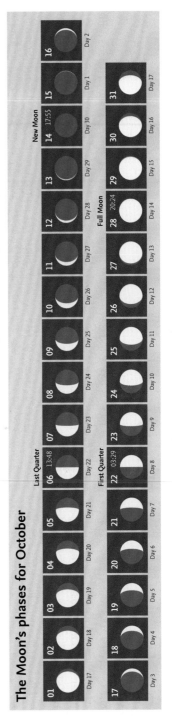

The constellation of Aquarius is one of the constellations that is visible in late summer and early autumn. The four stars forming the 'Y'-shape of the 'Water Jar' may be seen to the east of Sadalmelik (α Aquari), the brightest star (top centre).

October – Moon and Planets

The Moon

On October 2, the Moon passes 3.4° north of *Jupiter* (mag. -2.8), retrograding in *Aries*. Later that day it is 2.9° north of *Uranus* (mag. 5.7). It then passes the *Pleiades* and on October 4 is 9.4° north of *Aldebaran*. By October 7, one day after Last Quarter, it is 1.4° south of *Pollux* in *Gemini*. On October 10, the waning crescent is 4.2° north of *Regulus* in *Leo*. Later the same day it is 6.5° north of brilliant *Venus* (mag. -4.6) also in Leo. On October 14, at New Moon, there is an annular solar eclipse (see pages 20–21). Later that day, the Moon is 2.4° north of *Spica* in twilight. By October 18, it is 0.8° north of *Antares* in *Scorpius*. On October 24, the Moon is 2.8° south of *Saturn* in *Aquarius*. By October 29, one day after Full, the Moon is 3.1° north of *Jupiter* and on October 30

it is 2.9° north of *Uranus*, both in *Aries*. The next day, October 31, the waning gibbous Moon passes 9.4° north of *Aldebaran*.

The planets

Mercury is lost in twilight close to the Sun, and is at superior conjunction on October 20. *Venus*, in *Leo*, is bright (mag. -4.7 to -4.4) in the evening sky and moving closer to the Sun. *Jupiter* is retrograding in *Aries* at mag. -2.8 to -2.9. *Saturn* is also retrograding in *Aquarius*. *Uranus* is still in *Aries* at mag. 5.8 and *Neptune* is slowly retrograding in *Pisces* at mag. 7.8.

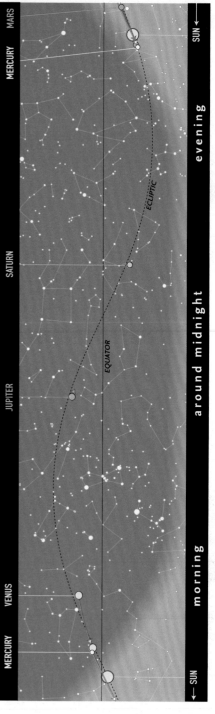

The path of the Sun and the planets along the ecliptic in October.

Calendar for October

02–Nov.07		Orionid meteor shower
02	03:20	Jupiter 3.4°S of the Moon
02	17:15	Uranus 2.9°S of the Moon
04	01:46	Aldebaran 9.4°S of the Moon
06–10		Draconid meteor shower
06	13:48	Last Quarter
07	11:02	Pollux 1.4°N of the Moon
08–09		Draconid meteor shower maximum
10–11		Southern Taurid meteor shower maximum
10	03:42	Moon at apogee = 405,426 km
10	09:20	Regulus 4.2°S of the Moon
14	09:44	Venus 6.5°S of the Moon
14	17:55	Mercury 0.7°N of the Moon
14	17:59	New Moon
14	22:07	Annular solar eclipse
15	16:17	Spica 2.4°S of the Moon
18	13:53	Mars 1.0°N of the Moon
20–Dec.10		Antares 0.8°S of the Moon
20	05:38	Northern Taurid meteor shower
21–22		Mercury at superior conjunction
		Orionid meteor shower maximum
22	03:29	First Quarter
23	23:14	Venus at greatest elongation (46.4°W, mag. –4.5)
24	07:56	Saturn 2.8°N of the Moon
26	01:23	Neptune 1.5°N of the Moon
26	03:02	Moon at perigee = 364,872 km
28	20:13	Partial lunar eclipse
28	20:24	Full Moon
29	01:00	European Daylight Saving Time (British Summer Time, BST) ends
29	08:14	Jupiter 3.1°S of the Moon
29	17:00 *	Mars (mag. 1.5) 0.4°N of Mercury (mag. –1.0)
30	01:53	Uranus 2.9°S of the Moon
31	11:28	Aldebaran 9.4°S of the Moon

These objects are close together for an extended period around this time.

Early morning 3:30 (BST)

Pleiades · Uranus · Jupiter

October 2–3 • *The Moon is high in the south, when it passes Jupiter, Uranus (mag. 5.7) and the Pleiades.*

Morning 6:00 (BST)

Castor · Pollux · Moon · Procyon · Alhena

October 7 • *The Moon is close to Pollux, high in the southeast. Castor, Alhena and Procyon are nearby.*

Morning 6:00 (BST)

Algieba · Regulus · Venus · Moon

October 10 • *The Moon lines up with Regulus and Venus, with Algieba to its left.*

Evening 22:00 (BST)

Saturn · Fomalhaut

October 23–24 • *The waxing gibbous Moon passes Saturn. Fomalhaut is closer to the horizon and almost due south.*

Evening 21:00 (22:00 BST)

Pleiades · Uranus · Jupiter · Aldebaran

October 28–30 • *The Moon passes Jupiter, Uranus (mag. 5.6) and the Pleiades. Aldebaran is ten degrees closer to the horizon.*

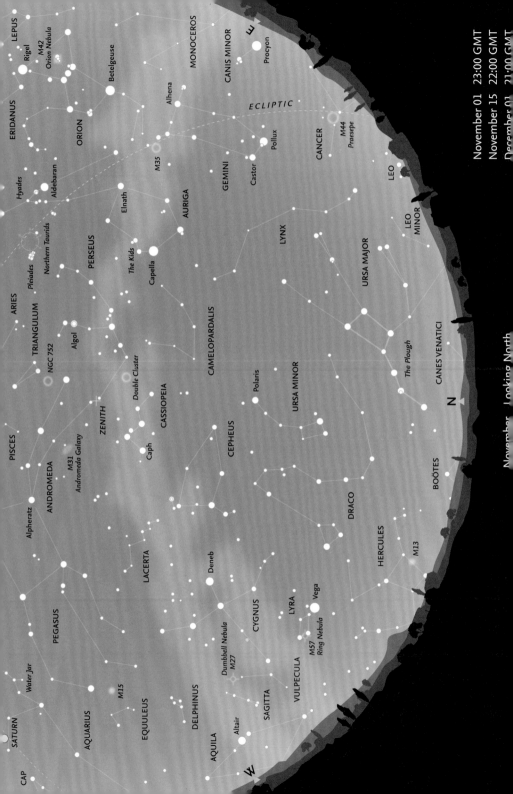

November Looking North

November – Looking North

The constellation of Auriga, with brilliant Capella (mag. 0.08), which, although appearing as a single star, is actually a quadruple system, consisting of a pair of yellow giant stars, gravitationally bound to a more distant pair of red dwarfs.

Most of **Aquila** has now disappeared below the horizon, but two of the stars of the 'Summer Triangle', **Vega** in **Lyra** and **Deneb** in **Cygnus**, are still clearly visible in the west. The head of **Draco** is now low in the northwest and only a small portion of **Hercules** remains above the horizon. The southernmost stars of **Ursa Major** are now coming into view. The Milky Way arches overhead, with the denser star clouds in the west and the less heavily populated region through **Auriga** and **Monoceros** in the east. High overhead, **Cassiopeia** is near the zenith and **Cepheus** has swung round to the northwest, while Auriga is now high in the northeast. **Gemini**, with **Castor** and **Pollux**, is well clear of the eastern horizon, and even **Procyon** (α Canis Minoris) is just climbing into view almost due east.

Meteors

The **Northern Taurid** shower, which began in mid-October, reaches maximum – although with just a low rate of about five meteors per hour – on November 12. New Moon is on November 13, so conditions are very favourable. The shower gradually trails off, ending around December 10. There is an apparent 7-year periodicity in fireball activity, but 2023 is unlikely to be a peak year. Far more striking, however, are the **Leonids**, which have a relatively short period of activity (November 6–30), with maximum on November 17–18. This shower is associated with Comet 55P/Tempel-Tuttle and has shown extraordinary activity on various occasions with many thousands of meteors per hour. High rates were seen in 1999, 2001 and 2002 (reaching about 3000 meteors per hour) but have fallen dramatically since then. The rate in 2023 is likely to be about 10 per hour. These meteors are the fastest shower meteors recorded (about 70 km per second) and often leave persistent trains. The shower is very rich in faint meteors. In 2023, maximum is when the Moon is a waxing crescent, so conditions are reasonably favourable.

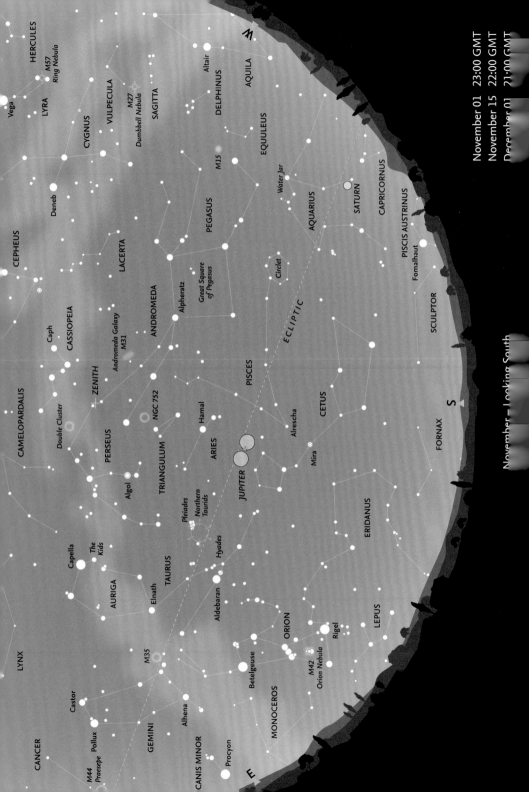

November 01 23:00 GMT
November 15 22:00 GMT
December 01 21:00 GMT

November – Looking South

November – Looking South

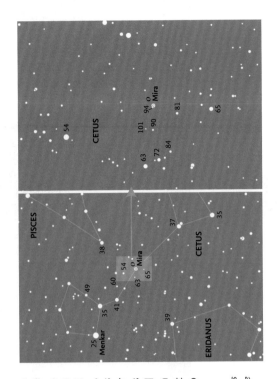

Orion has now risen above the eastern horizon, and part of the long, straggling constellation of **Eridanus** (which begins near **Rigel**) is visible to the west of Orion. Higher in the sky, **Taurus**, with the **Pleiades** cluster, and orange **Aldebaran** are now easy to observe. To their west, both **Pisces** and **Cetus** are close to the meridian. The famous long-period variable star, **Mira** (o Ceti), with a typical range of magnitude 3.4 to 9.8, is favourably placed for observation. In the southwest, **Capricornus** has slipped below the horizon, but **Aquarius** remains visible. Even farther west, **Altair** may be seen early in the night, but most of **Aquila** has already disappeared from view. **Delphinus**, together with **Sagitta** and **Vulpecula** in the Milky Way, will soon vanish for another year. Both **Pegasus** and **Andromeda** are easy to see, and one of the lines of stars that make up Andromeda finishes close to the zenith, which is also close to one of the outlying stars of **Perseus**, high in the east.

Finder and comparison charts for Mira (o Ceti). The chart on the left shows all stars brighter than magnitude 6.5. The chart on the right shows stars down to magnitude 10.0. The comparison star magnitudes are shown without the decimal point.

The Moon's phases for November

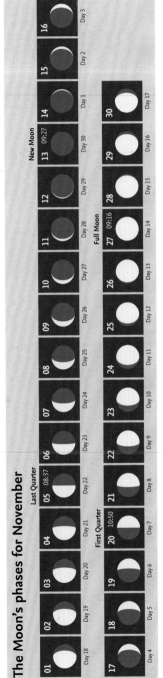

November – Moon and Planets

The Moon

On November 3, the waning gibbous Moon is 1.5° south of *Pollux*. On November 6 it passes 4.2° north of *Regulus*, between it and *Algieba*. By November 9 it is 1.0° north of *Venus* (mag. -4.4) in *Virgo*. On November 11 it passes 2.4° north of *Spica* in *Virgo*. By November 20, the Moon is 2.7° south of *Saturn* in *Aquarius*. On November 25 it passes 2.8° north of *Jupiter* in *Aries*, then the next day, 2.6° north of *Uranus* (mag. 5.6) also in *Aries*. On November 27 it is 9.3° north of *Aldebaran*.

The Planets

Mercury is in evening twilight, but may become visible at the end of the month on the border of *Ophiuchus* and *Sagittarius* at mag. -0.4 to -0.5. *Venus* (mag. -4.4 to -4.2) begins in *Leo* and moves across *Virgo*. *Mars* is lost in twilight near the Sun. *Jupiter* (mag. -2.9) comes to opposition in *Aries* on November 3. *Saturn* (mag. 0.7 to 0.8) initially retrograding, resumes direct motion on November 5. *Uranus* (mag. 5.6) is at opposition in *Aries* on November 13 and *Neptune* is slowly retrograding in *Pisces* at mag. 7.9.

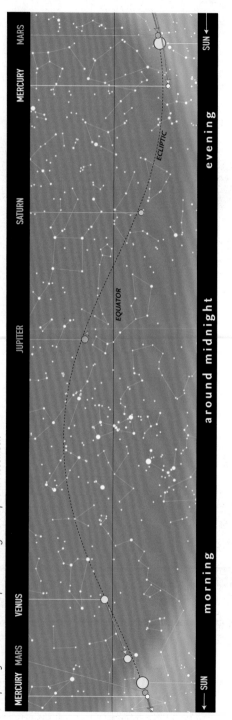

The path of the Sun and the planets along the ecliptic in November.

03	05:02	Jupiter at opposition (mag. –2.9)
03	19:09	Pollux 1.5°N of the Moon
05	08:37	Last Quarter
06–30		Leonid meteor shower
06	16:59	Regulus 4.2°S of the Moon
06	21:49	Moon at apogee = 404,569 km
09	09:30	Venus 1.0°S of the Moon
1	05:48	Spica 2.4°S of the Moon
2–13		Northern Taurid meteor shower maximum
3	09:27	New Moon
3	13:32	Mars 2.5°N of the Moon
3	17:21	Uranus at opposition (mag. 5.6)
4	14:39	Mercury 1.7°N of the Moon
4	20:18	Antares 0.9°S of the Moon
7–18		Leonid meteor shower maximum
8	05:41	Mars at superior conjunction
0	10:50	First Quarter
0	14:06	Saturn 2.7°N of the Moon
1	21:01	Moon at perigee = 369,818 km
2	07:45	Neptune 1.5°N of the Moon
5	11:14	Jupiter 2.8°S of the Moon
6	09:19	Uranus 2.6°S of the Moon
7	09:16	Full Moon
7	21:03	Aldebaran 9.2°S of the Moon

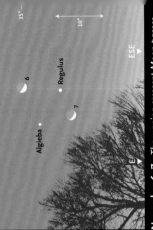

Evening 22:00

November 3 • The Moon is in the company of Pollux and Castor, when it rises in the northeast. Alhena is farther east.

Early morning 3:00

November 6–7 • The waning crescent Moon passes between Regulus and Algieba.

Morning 6:00

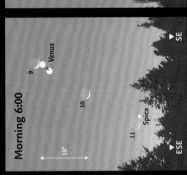

November 9–11 • The Moon is close to Venus on November 9. Two days later it passes Spica.

Evening 17:30

November 20 • The Moon is near Saturn. A line through the two objects leads to Fomalhaut.

Early morning 3:00

November 25–27 • The Moon passes Jupiter, Uranus (mag. 5.6) and the Pleiades. Aldebaran is nearby.

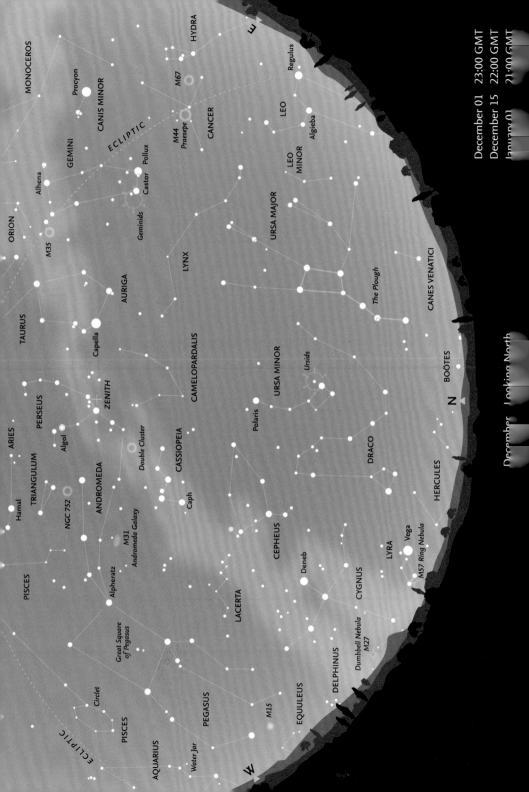

December 01   23:00 GMT
December 15   22:00 GMT
January 01   21:00 GMT

December   Looking North

December – Looking North

Ursa Major has now swung around and is starting to 'climb' in the east. The fainter stars in the southern part of the constellation are now fully in view. The other bear, *Ursa Minor*, 'hangs' below *Polaris* in the north. Directly above it is the faint constellation of *Camelopardalis*, with the other inconspicuous circumpolar constellation, *Lynx*, to its east. *Vega* (α Lyrae) is skimming the horizon in the northwest, but *Deneb* (α Cygni) and most of *Cygnus* remain visible farther west. In the east, *Regulus* (α Leonis) and the constellation of *Leo* are beginning to rise above the horizon. *Cancer* stands high in the east, with *Gemini* even higher in the sky. *Perseus* is at the zenith, with *Auriga* and *Capella* between it and Gemini. Because it is so high in the sky, now is a good time to examine the star clouds of the fainter portion of the Milky Way, between *Cassiopeia* in the west to Gemini and *Orion* in the east.

Meteors

There is one significant meteor shower in December (the last major shower of the year). This is the *Geminid* shower, which is visible over the period December 4–20 and comes to maximum on December 14–15, when the Moon is waxing crescent, so conditions are favourable. It is one of the most active showers of the year, and in some years is the strongest, with a peak rate of around 100 meteors per hour. It is the one major shower that shows good activity before midnight. The meteors have been found to have a much higher density than other meteors (which are derived from cometary material). It was eventually established that the Geminids and the asteroid Phaethon had similar orbits. So the Geminids are assumed to consist of denser, rocky material. They are slower than most other meteors and often appear to last longer. The brightest often break up into numerous luminous fragments that follow similar paths across the sky. There is a second shower: the *Ursids*, active December 17–26, peaking on December 22–23,

with a rate at maximum of 5–10, occasionally rising to 25 per hour. Maximum in 2023 occurs when the Moon is waxing gibbous, so conditions are poor. The parent body is Comet 8P/Tuttle.

The constellation of Andromeda largely consists of a line of bright stars running northeast from α Andromedae, Alpheratz (bottom right), which is one of the stars forming the Great Square of Pegasus. The small constellation of Triangulum appears on the left-hand (eastern) side, below Andromeda.

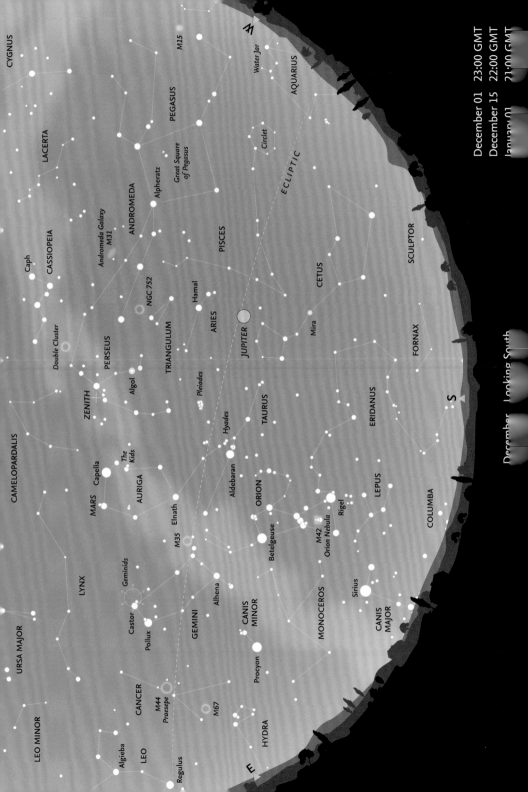

December · Looking South

December – Looking South

The fine open cluster of the **Pleiades** is due south around 22:00, with the **Hyades** cluster, **Aldebaran** and the rest of **Taurus** clearly visible to the east. **Auriga** (with **Capella**) and **Gemini** (with **Castor** and **Pollux**) are both well-placed for observation. **Orion** has made a welcome return to the winter sky, and both **Canis Minor** (with **Procyon**) and **Canis Major** (with **Sirius**, the brightest star in the sky) are now well above the horizon. The small, poorly known constellation of **Lepus** lies to the south of Orion. In the west, **Aquarius** has now disappeared, and **Cetus** is becoming lower, but **Pisces** is still easily seen, as are the constellations of **Aries**, **Triangulum** and **Andromeda** above it. The Great Square of **Pegasus** is starting to plunge down towards the western horizon, and because of its orientation on the sky, appears more like a large diamond, standing on one point, than a square.

The constellation of Taurus contains two contrasting open clusters: the compact Pleiades, with its striking blue-white stars, and the more scattered, 'V'-shaped Hyades, which are much closer to us. Orange Aldebaran (α Tauri) is not related to the Hyades, but lies between it and the Earth.

The Moon's phases for December

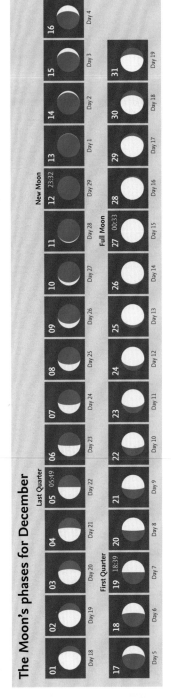

December – Moon and Planets

The Moon

On December 1, the Moon is 1.6° south of *Pollux* in *Gemini*. On December 4 the Moon passes 4.0° north of *Regulus* between it and *Algieba*. By December 8 the waning crescent Moon is 2.3° north of *Spica*. The next day it is 3.6° south of *Venus*. By New Moon on December 12, lost in twilight, it is 0.9° north of *Antares* and later that day, 3.6° south of *Mars*. On December 17 the Moon is 2.5° south of *Saturn* in *Aquarius*. By December 22 the waxing gibbous Moon is 2.6° north of *Jupiter* in *Aries*. It is 2.8° north of Uranus (mag. 5.7), also in Aries, the next day. It passes 9.4° north of *Aldebaran* on December 25. One day after Full Moon, on December 28, it is 1.7° south of *Pollux* and on December 31 3.8° north of *Regulus*, again between it and *Algieba*.

The planets

Mercury is at greatest eastern elongation (21.3°) on December 4 at mag -0.5. *Venus* (mag. -4.1) moves from *Virgo* into *Libra* and morning twilight. *Mars* remains too close to the Sun to be visible. *Jupiter* (mag. -2.8 to -2.6) is in *Aries*. *Saturn* is moving slowly eastwards in *Aquarius*. *Uranus* is in *Aries* and *Neptune* is still slowly retrograding in Pisces at mag. 7.9. The minor planet **(4)** *Vesta* is at opposition (mag. 6.4) on December 21 (see page 26) and **(9)** *Metis* (mag. 8.4) at opposition the next day in *Gemini*.

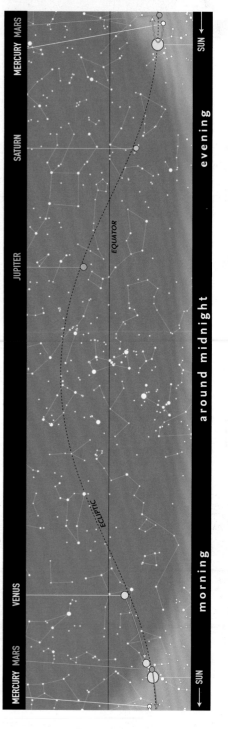

The path of the Sun and the planets along the ecliptic in December.

Calendar for December

01	04:01	Pollux 1.6°N of the Moon
04–20		Geminid meteor shower
04	01:18	Regulus 4.0°S of the Moon
04	14:28	Mercury at greatest elongation (17.9°E, mag. -0.5)
04	18:42	Moon at apogee = 404,346 km
05	05:49	Last Quarter
08	14:45	Spica 2.3°S of the Moon
09	16:53	Venus 3.6°N of the Moon
12	04:52	Antares 0.9°S of the Moon
12	10:55	Mars 3.6°N of the Moon
12	23:32	New Moon
14–15		Geminid meteor shower maximum
14	05:19	Mercury 4.4°N of the Moon
16	18:53	Moon at perigee = 367,901 km
17–26		Ursid meteor shower
17	22:01	Saturn 2.5°N of the Moon
19	13:16	Neptune 1.3°N of the Moon
19	18:39	First Quarter
21	18:56	Minor planet (4) Vesta at opposition (mag. 6.4)
22–23		Ursid meteor shower maximum
22	03:27	Winter solstice
22	14:24	Jupiter 2.6°S of the Moon
22	18:54	Mercury at inferior conjunction
22	23:26	Minor planet (9) Metis at opposition (mag. 8.4)
23	14:54	Uranus 2.8°S of the Moon
25	05:01	Aldebaran 9.4°S of the Moon
27	00:33	Full Moon
28	03:00 *	Mercury (mag. 2.0) 3.6°N of Mars (mag. 1.4)
28	12:30	Pollux 1.7°N of the Moon
31	09:33	Regulus 3.8°S of the Moon

* These objects are close together for an extended period around this time.

Morning 7:00

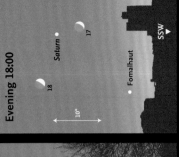

December 1 • The Moon is lining up with Pollux and Castor, shortly before sunrise.

Evening 17:00

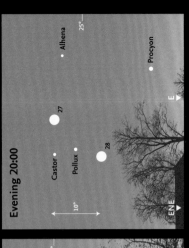

December 22–24 • The Moon passes Jupiter, Uranus (mag. 5.7) and the Pleiades. On December 24, Aldebaran is ten degrees below the Moon.

Morning 7:00

December 8–9 • In the morning twilight, the Moon passes Spica and the very bright Venus (mag. -4.0).

Evening 18:00

December 17–18 • The Moon passes between Saturn (mag. 1.1) and Fomalhaut (mag. 1.2).

Evening 20:00

December 27–28 • In the east, the Full Moon passes below Pollux and Castor. Procyon and Alhena are close by.

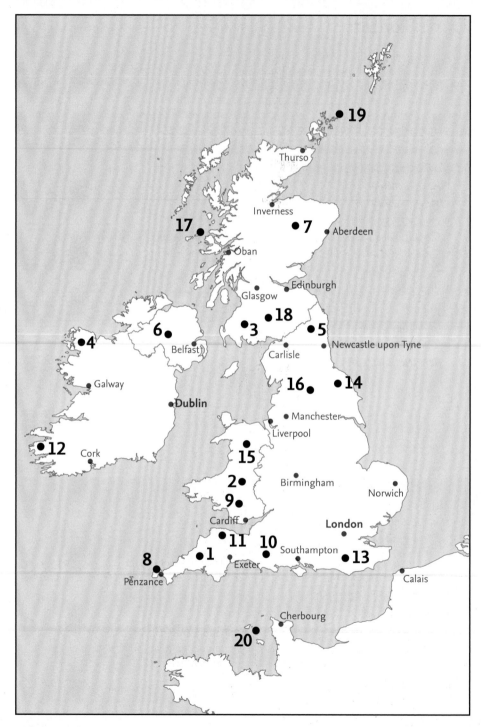

Dark Sky Sites

International Dark-Sky Association Sites

The **International Dark-Sky Association** (IDA) recognizes various categories of sites that offer areas where the sky is dark at night, free from light pollution and particularly suitable for astonomical observing. A number of sites in Great Britain and Ireland have been given specific recognition and are shown on the map. These are:

Parks

1 *Bodmin Moor Dark Sky Landscape*
2 *Elan Valley Estate*
3 *Galloway Forest Park*
4 *Mayo Dark Sky Park*
5 *Northumberland National Park and Kielder Water & Forest Park*
6 *OM Dark Sky Park & Observatory*
7 *Tomintoul and Glenlivet – Cairngorms*
8 *West Penwith*

Reserves

9   *Brecon Beacons National Park*
10 *Cranborne Chase*
11 *Exmoor National Park*
12 *Kerry*
13 *Moore's Reserve South Down National Park*
14 *North York Moors National Park*
15 *Snowdonia National Park*
16 *Yorkshire Dales National Park*

Communities

17 *The island of Coll (Inner Hebrides, Scotland)*
18 *Moffat*
19 *North Ronaldsay Dark Sky Island (Orkney Islands, Scotland)*
20 *The island of Sark (Channel Islands)*

Details of these sites and web links may be found at the IDA website: https://www.darksky.org/ Many of these sites have major observatories or other facilities available for public observing (often at specific dates or times).

Dark Sky Discovery Sites

In Britain there is also the **Dark Sky Discovery** organisation. This gives recognition to smaller sites, again free from immediate light pollution, that are open to observing at any time. Some sites are used for specific, public observing sessions. A full listing of sites is at:

https://www.darkskydiscovery.org.uk/

but specific events are publicized locally.

Glossary and Tables

aphelion   The point on an orbit that is farthest from the Sun.

apogee   The point on its orbit at which the Moon is farthest from the Earth.

appulse   The apparently close approach of two celestial objects; two planets, or a planet and star.

astronomical unit   (AU) The mean distance of the Earth from the Sun, 149,597,870 km.

celestial equator   The great circle on the celestial sphere that is in the same plane as the Earth's equator.

celestial sphere   The apparent sphere surrounding the Earth on which all celestial bodies (stars, planets, etc.) seem to be located.

conjunction   The point in time when two celestial objects have the same celestial longitude. In the case of the Sun and a planet, superior conjunction occurs when the planet lies on the far side of the Sun (as seen from Earth). For Mercury and Venus, inferior conjuction occurs when they pass between the Sun and the Earth.

direct motion   Motion from west to east on the sky.

ecliptic   The apparent path of the Sun across the sky throughout the year. Also: the plane of the Earth's orbit in space.

elongation   The point at which an inferior planet has the greatest angular distance from the Sun, as seen from Earth.

equinox   The two points during the year when night and day have equal duration. Also: the points on the sky at which the ecliptic intersects the celestial equator. The vernal (spring) equinox is of particular importance in astronomy.

gibbous   The stage in the sequence of phases at which the illumination of a body lies between half and full. In the case of the Moon, the term is applied to phases between First Quarter and Full, and between Full and Last Quarter.

inferior planet   Either of the planets Mercury or Venus, which have orbits inside that of the Earth.

magnitude   The brightness of a star, planet or other celestial body. It is a logarithmic scale, where larger numbers indicate fainter brightness. A difference of 5 in magnitude indicates a difference of 100 in actual brightness, thus a first-magnitude star is 100 times as bright as one of sixth magnitude.

meridian   The great circle passing through the North and South Poles of a body and the observer's position; or the corresponding great circle on the celestial sphere that passes through the North and South Celestial Poles and also through the observer's zenith.

nadir   The point on the celestial sphere directly beneath the observer's feet, opposite the zenith.

occultation   The disappearance of one celestial body behind another, such as when stars or planets are hidden behind the Moon.

opposition   The point on a superior planet's orbit at which it is directly opposite the Sun in the sky.

perigee   The point on its orbit at which the Moon is closest to the Earth.

perihelion   The point on an orbit that is closest to the Sun.

retrograde motion   Motion from east to west on the sky.

superior planet   A planet that has an orbit outside that of the Earth.

vernal equinox   The point at which the Sun, in its apparent motion along the ecliptic, crosses the celestial equator from south to north. Also known as the First Point of Aries.

zenith   The point directly above the observer's head.

zodiac   A band, stretching 8° on either side of the ecliptic, within which the Moon and planets appear to move. It consists of twelve equal areas, originally named after the constellation that once lay within it.

The Greek Alphabet

α	Alpha	ε	Epsilon	ι	Iota	ν	Nu	ρ	Rho	φ (φ)	Phi
β	Beta	ζ	Zeta	κ	Kappa	ξ	Xi	σ (ς)	Sigma	χ	Chi
γ	Gamma	η	Eta	λ	Lambda	ο	Omicron	τ	Tau	ψ	Psi
δ	Delta	θ (ϑ)	Theta	μ	Mu	π	Pi	υ	Upsilon	ω	Omega

The Constellations

There are 88 constellations covering the whole of the celestial sphere, but 24 of these in the southern hemisphere can never be seen (even in part) from the latitude of Britain and Ireland, so are omitted from this table. The names themselves are expressed in Latin, and the names of stars are frequently given by Greek letters followed by the genitive of the constellation name. The genitives and English names of the various constellations are included.

Name	Genitive	Abbr.	English name
Andromeda	Andromedae	And	Andromeda
Antlia	Antliae	Ant	Air Pump
Aquarius	Aquarii	Aqr	Water Bearer
Aquila	Aquilae	Aql	Eagle
Aries	Arietis	Ari	Ram
Auriga	Aurigae	Aur	Charioteer
Boötes	Boötis	Boo	Herdsman
Camelopardalis	Camelopardalis	Cam	Giraffe
Cancer	Cancri	Cnc	Crab
Canes Venatici	Canum Venaticorum	CVn	Hunting Dogs
Canis Major	Canis Majoris	CMa	Big Dog
Canis Minor	Canis Minoris	CMi	Little Dog
Capricornus	Capricorni	Cap	Sea Goat
Cassiopeia	Cassiopeiae	Cas	Cassiopeia
Centaurus	Centauri	Cen	Centaur
Cepheus	Cephei	Cep	Cepheus
Cetus	Ceti	Cet	Whale
Columba	Columbae	Col	Dove
Coma Berenices	Comae Berenices	Com	Berenice's Hair
Corona Australis	Coronae Australis	CrA	Southern Crown
Corona Borealis	Coronae Borealis	CrB	Northern Crown
Corvus	Corvi	Crv	Crow
Crater	Crateris	Crt	Cup
Cygnus	Cygni	Cyg	Swan
Delphinus	Delphini	Del	Dolphin
Draco	Draconis	Dra	Dragon
Equuleus	Equulei	Equ	Little Horse
Eridanus	Eridani	Eri	River Eridanus
Fornax	Fornacis	For	Furnace
Gemini	Geminorum	Gem	Twins
Hercules	Herculis	Her	Hercules
Hydra	Hydrae	Hya	Water Snake

Name	Genitive	Abbr.	English name
Lacerta	Lacertae	Lac	Lizard
Leo	Leonis	Leo	Lion
Leo Minor	Leonis Minoris	LMi	Little Lion
Lepus	Leporis	Lep	Hare
Libra	Librae	Lib	Scales
Lupus	Lupi	Lup	Wolf
Lynx	Lyncis	Lyn	Lynx
Lyra	Lyrae	Lyr	Lyre
Microscopium	Microscopii	Mic	Microscope
Monoceros	Monocerotis	Mon	Unicorn
Ophiuchus	Ophiuchi	Oph	Serpent Bearer
Orion	Orionis	Ori	Orion
Pegasus	Pegasi	Peg	Pegasus
Perseus	Persei	Per	Perseus
Pisces	Piscium	Psc	Fishes
Piscis Austrinus	Piscis Austrini	PsA	Southern Fish
Puppis	Puppis	Pup	Stern
Pyxis	Pyxidis	Pyx	Compass
Sagitta	Sagittae	Sge	Arrow
Sagittarius	Sagittarii	Sgr	Archer
Scorpius	Scorpii	Sco	Scorpion
Sculptor	Sculptoris	Scl	Sculptor
Scutum	Scuti	Sct	Shield
Serpens	Serpentis	Ser	Serpent
Sextans	Sextantis	Sex	Sextant
Taurus	Tauri	Tau	Bull
Triangulum	Trianguli	Tri	Triangle
Ursa Major	Ursae Majoris	UMa	Great Bear
Ursa Minor	Ursae Minoris	UMi	Lesser Bear
Vela	Velorum	Vel	Sails
Virgo	Virginis	Vir	Virgin
Vulpecula	Vulpeculae	Vul	Fox

Some common asterisms

Belt of Orion	δ, ε and ζ Orionis
Big Dipper	α, β, γ, δ, ε, ζ and η Ursae Majoris
Circlet	γ, θ, ι, λ and κ Piscium
Guards (or Guardians)	β and γ Ursae Minoris
Head of Cetus	α, γ, ξ², μ and λ Ceti
Head of Draco	β, γ, ξ and ν Draconis
Head of Hydra	δ, ε, ζ, η, ρ and σ Hydrae
Keystone	ε, ζ, η and π Herculis
Kids	ζ and η Aurigae
Little Dipper	β, γ, η, ζ, ε, δ and α Ursae Minoris
Lozenge	= Head of Draco
Milk Dipper	ζ, γ, σ, φ and λ Sagittarii
Plough or Big Dipper	α, β, γ, δ, ε, ζ and η Ursae Majoris
Pointers	α and β Ursae Majoris
Sickle	α, η, γ, ζ, μ and ε Leonis
Square of Pegasus	α, β and γ Pegasi with α Andromedae
Sword of Orion	θ and ι Orionis
Teapot	γ, ε, δ, λ, φ, σ, τ and ζ Sagittarii
Wain (or Charles' Wain)	= Plough
Water Jar	γ, η, κ and ζ Aquarii
Y of Aquarius	= Water Jar

Acknowledgements

Denis Buczynski, Portmahomack, Ross-shire – p.32 (Fireball), p.35 (Quadrantid fireball)
Stephen Edberg – pp.35, 53, 89, 91, 103 (Constellation photographs)
Akira Fuji – p.21 (Lunar eclipse)
Jens Hackmann, Bad Mergentheim, Germany – p.77 (Perseid fireball)
Bernhard Hubl – pp. 49, 55, 71, 85, 95 (Constellation photographs)
Nick James, p.29 (Comet NEOWISE)
Damian Peach, p.23 (Mars photo)
peresanz/Shutterstock – p.37 (Orion)
Ken Sperber, California – p.83 (Double Cluster)
Wil Tirion, Capelle aan den IJssel, The Netherlands, p.43, 67, 101 (Constellation photographs)
Alan Tough, Nairn, Highland, Scotland, p.65 (noctilucent clouds)
Specialist editorial support was provided by Dr Gregory Brown – Public Astronomy Officer at the Royal
Observatory, part of Royal Museums Greenwich.